THE SUSTAINABLE CITY

—

STEVEN COHEN

THE
SUSTAINABLE
CITY

COLUMBIA UNIVERSITY PRESS
NEW YORK

Columbia University Press
Publishers Since 1893
New York Chichester, West Sussex
cup.columbia.edu
Copyright © 2018 Columbia University Press

Library of Congress Cataloging-in-Publication Data
Names: Cohen, Steven, 1953 September 6– author.
Title: The sustainable city / Steven Cohen.
Description: New York : Columbia University Press, [2017] | Includes
bibliographical references and index.
Identifiers: LCCN 2017012419 (print) | LCCN 2017030512 (ebook) |
ISBN 9780231543972 (electronic) | ISBN 9780231182041 (cloth : alk. paper) |
ISBN 9780231182058 (pbk.)
Subjects: LCSH: City planning—Environmental aspects. | Urban renewal. |
Sustainable development.
Classification: LCC HT166 (ebook) | LCC HT166 .C5284 2017 (print) | DDC
307.3/416—dc23
LC record available at https://lccn.loc.gov/2017012419

Columbia University Press books are printed on permanent
and durable acid-free paper.
Printed in the United States of America

Cover design: Julia Kushnirsky
Cover image: © Artem Vorobiev/Getty Images

To the memory of my mother, Shirley Balas Cohen.

CONTENTS

PART II. CASES IN URBAN SUSTAINABILITY

—

PART III. CONCLUSIONS

—

PREFACE

A great paradox of the transition to a sustainable economy is that it will not be achieved in rural places in harmony with nature but in cities built to exploit nature without destroying it. The urban migration now under way is a worldwide phenomenon that reached a critical inflection point in 2007, when for the first time most of the people on the planet lived in cities.

This book focuses on cities because place matters. Economic, technological, and cultural forces are moving people out of rural areas and into urban areas. While the global economy leads to a homogenization of fashion, entertainment, and aspects of culture, the human need for a sense of place and distinctiveness is countering some of these trends. But for cities to be sustainable in the brain-based economy, we must secure public investment in infrastructure, education, health care, and social services. We need to invest time, energy, and money in creating the sustainable city.

The infrastructure to support distributed generation of renewable energy, mass and personal transit, and to treat and transform water, sewage, and solid waste can be built and even managed by private contractors but still requires a public sector that is active, ethical, sophisticated, and able to form productive public-private partnerships. Many cities are already investing in different parts of the infrastructure of the future, but the modern city still has a long way to go. This book provides a broad overview of the sustainable city from an organizational management and public policy perspective, utilizing examples and case studies from initiatives, projects, policies, and legislation already extant in cities around the world. It examines past trends and potential future ones, contains real solutions and applications, and looks at the key aspects of a sustainable urban lifestyle. The book is divided into three parts: I, Concepts; II, Cases; and III, Conclusions.

The book begins by defining the sustainable city, at least for the purposes of the discussion I will present. Why should a book on sustainability focus on cities? Because cities provide the dynamic, social, and ever-changing environment that people thrive in. We want to be part of the center of activity, to feel that we "belong" and yet that we are still distinctive. It is in the world's cities that human potential can be realized—we just need to make sure we don't destroy the planet while we explore that potential. While our economic life allows us to pay less attention to our basic biological needs, those needs remain and cannot be met if our planet is toxic or dangerous. We can have urban dynamism and clean air, water, and food, but it requires competent sustainability management, good governance, and adequate financial resources to be accomplished. The desire for a clean and healthy environment is an inevitable stage of economic development. The tainted water supply in Flint, Michigan, and the ensuing political crisis in 2016 for Michigan's governor is only a small example of how a failure of governance can poison people. We know how to process and deliver clean water. The technology to transform even wastewater to safe drinking water exists and is available. But it is not free.

What are the elements of a sustainable city? The overall definition of such a city is one that facilitates human economic (production and consumption) and social life with the least possible impact on the natural environment. This means that material flows into and out of the city are thought through and managed to minimize destruction of natural systems. The city's water supply comes from sources that are replenished through natural processes or from sources that can be withdrawn without damaging ecosystems. The city's solid waste is recycled as much as possible with food waste converted to fertilizer and other materials separated for reuse as well. Other systems such as sewage treatment, storm water drainage, energy, food, and transportation are designed for efficiency and least possible environmental impact.

The sustainable city sounds like a wonderful place to live, but how do we get from here to there? How do we make the transition from today's unsustainable city to tomorrow's sustainable one? There are three arenas that must be activated for this transition to take place: (1) organizational management, (2) finance, and (3) public policy. Chapter 1 of this book provides an overview of the type of change that is needed in each of these arenas.

Chapter 2 defines and explains sustainable urban systems. I define and explain the following urban systems and what "sustainability" means in each of these distinct systems:

- Energy
- Water
- Waste
- Sewage
- Food
- Transport
- Public space

I also define and assess the technical, financial, organizational, and political requirements of the sustainable city. Each of these sections could easily be a book itself, and so by necessity this chapter can only provide an overview.

Chapter 3 discusses the sustainable urban lifestyle. My effort here is to distinguish the way people live in a sustainable city from that of people who attempt to live close to nature in rural areas. Obviously, urbanites pursuing a sustainable lifestyle are not living off the grid, growing all their food, and disposing their food waste in a compost heap. However, they may well grow some food in a neighborhood garden, participate in a farm-share where they guarantee they will purchase the produce of a local farmer, use renewable energy, practice energy efficiency, and send their food waste to an anaerobic digester.

The sustainable city involves a sustainable lifestyle and a transformation from the consumer society to something else. The twenty-first century brain-based economy has changed the nature of production and consumption. Rather than being defined by the size of one's home and the consumer items one possesses, the sustainable lifestyle involves a search for different values. Consumer items remain valued, but they become means rather than ends. Consumption becomes more oriented toward services, entertainment, travel, and experiences and less oriented toward possessing manufactured products. Such products become commodities in the sustainable city, providing the necessities of life, but no longer serving as self-justifying goals. Culture and values are far more powerful forces of social change and consumption patterns than regulation. This chapter defines and explains the sustainable urban lifestyle that can be achieved in the sustainable city. It includes discussion of

- the changing nature of consumption;
- fitness, wellness, longevity, and health care;
- lifelong learning;
- the role and importance of place;
- the changing nature of work; and
- the connected world culture.

In many respects this will be the most speculative chapter of the book. I really can't predict the future, and even though I see this new lifestyle emerging, unpredictable events such as climate change, terrorism, or economic depression could easily undercut these projections.

Chapter 4 leaves sociology behind to focus on organizational management and the movement toward an organizational focus on the physical dimensions of sustainability. Today's corporations, nonprofits, and governments are operating on a more crowded and interconnected planet that provides great opportunities but also poses great threats. There is little question that organizations are responding to greater population, consumption, resource scarcity, environmental degradation, and increased risk and liability and factoring these issues into routine decision making. Energy, water, and other raw materials are becoming a larger element of the cost structure of all organizations. Organizations are now routinely looking at energy consumption as a way to cut costs and increase efficiency. Global competition means that someone someplace is doing the same thing you are, and if you don't keep improving they could knock you out of business. Similar trends can be seen as organizations assess their use of water and other material resources. The costs and impact of waste, discharges of effluents, and emissions are now subject to critical analysis. This is not only for regulatory compliance:

in some organizations, the risk of environmental effects and the cost of insuring those risks have become part of routine decision making. This chapter discusses these issues and provides examples of organizations that now include the physical dimension of sustainability as basic management inputs.

The book then turns to a discussion of the role of politics and public policy in building sustainable cities. We needed public-private partnerships to build the industrial city of the nineteenth and twentieth centuries. Government and private corporations worked together to attract the capital and expertise required to build the energy, water, sewage, transport, and other infrastructure needed to ensure that people and businesses could live in thriving but dense modern settlements. We once again require such partnerships to build the sustainable city of the twenty-first century. While in the United States we desperately need federal sustainability policy, in the final analysis the environmental quality that people experience in their home communities will have the highest degree of political salience. A successful strategy to protect our environment will need to focus on local effects.

Chapter 5 addresses the role of sustainability in local and state politics, particularly in the United States, and examines the potential role of local and state governments in developing the public policies and partnerships needed to build sustainable infrastructure. I will examine local initiatives outside the United States as well, but the emphasis will be on the United States. The chapter addresses these issues:

- What is the role of sustainability in local politics?
    Mobilizing voters and building consensus: Can sustainability be nonpartisan?
    Public opinion and values related to sustainability.
    The challenge and opportunity of the NIMBY ("not in my backyard") syndrome.
    The link of environment to health.
- Can sustainability integrate economic and environmental goals and bring environmental and business interests together?
    What goals are shared?
    Is there a basis for compromise?

This chapter examines sustainability politics at the local level and details the force of these issues today and their potential in the future. Some issues are so inherently local that they require action at the city level. This chapter identifies the sustainability issues that are typically subject to local discretion and initiative both inside and outside the United States.

After defining and analyzing the concept of the sustainable city, I then illustrate the concepts I've introduced with a set of case studies in part II of the book. The goal of the case studies is to provide detailed examples of urban sustainability programs, policies, and projects that are in place and can be assessed. The concepts delineated and the issues raised in part I are illuminated and brought to life by the cases. The cases provide concrete examples of the actions needed to transition to sustainable cities.

Chapters 6 through 10 include the following case studies:

Waste management: New York, Hong Kong, and Beijing
Transportation: Bus rapid transit in Bogotá, light rail in Jerusalem, high-speed rail in China
Energy infrastructure: Microgrids and smart grids in New York, Japan, and Africa
Public space: The High Line, Gas Works Park, Victor Civita Plaza, and Canal Park
Sustainable urban living and the sharing economy: Uber and Airbnb

The book concludes in part III with a summary of the policy, management, and political lessons learned throughout the book. The book's conclusion also frankly discusses uncertainties and issues that require additional research. While I am confident that the transition to a sustainable and renewable economy will take place in the world's cities, I am far from confident that I understand how that change will take place. My hope is that this volume will provide the basis for further discussion, research, and analysis of the transition to sustainable cities.

ACKNOWLEDGMENTS

As always, everything I work on is a team effort. Alison Miller, Hayley Martinez, Kelsie DeFrancia, and Alix Schroder were the key staff who helped me turn these ideas into an integrated essay. A number of research assistants, all students at Columbia University, worked on the cases and source materials for this book. I thank Hilary Osborn and Catalina Villegas for their time spent researching and editing drafts. I rely on the excellence of the staff and students who worked with me on this book, and I am grateful for their hard work and brainpower.

A number of colleagues were critical in framing many of the ideas I write about in this book. As always, I am grateful to my friend, colleague, and longtime professional partner, Bill Eimicke. Other faculty who also were critical include Howard Apsan, Satyajit Bose, Howard Buffet, Mark Cane, Peter Coleman, Ruth DeFries, Peter DeMenocal, David Dinkins, Ester Fuchs, Michael Gerrard, Dong Guo, Tanya Heikkila, Sheldon Kamieniecki, Upmanu Lall, Peter Marcotullio, Vijay Modi, Kate Orff, Richard Plunz, Curtis Probst, Louise Rosen, Jeffrey Sachs, George Sarrinikolaou, Elliott Sclar, Sean Solomon, Lynnette Widder and too many others to name here. My great mentor and friend, Marc Tipermas, continues to influence my ideas about organizations and the importance of agile management. Marc along with Tom Brisbin of the Willdan Group have been teaching me key lessons of the importance of the private sector in building a sustainable nation. I am grateful for the example of Lee Bollinger, Columbia University's visionary president, who shows all of us how to combine thought leadership with organizational leadership.

I am profoundly grateful for the deep and loving support of my wife, Donna Fishman, and of my daughters, Gabriella Rose Cohen and Ariel Mariah Cohen. I thank my brother, Robby, my sisters, Judith and Myra, and my dad, Marvin Cohen. I am dedicating this book to my late mother, Shirley Balas Cohen. My mother could and would talk to anyone about anything. She was one of a kind. I don't know if she would have enjoyed this book, but she would have read it and then let me know how it could have been improved. I miss the sharpness and the softness of my mother's voice. People tell me that when I write I have a voice they can recognize. A lot of that voice was her voice. Her generation of women made it possible for my daughters to grow up assuming feminist values as a given. While she did not directly benefit from her struggle, the rest of us did. I want my children to live in a sustainable world that prizes peace over avarice and trust over fear. This is a time in America when we must work even harder to build a peaceful, sustainable nation and planet. I hope this book is a small contribution to the creation of that world.

THE SUSTAINABLE CITY

CONCEPTS

1

DEFINING THE SUSTAINABLE CITY

A DEFINITION

Because a city is a human settlement that is designed for human rather than ecological well-being, it may seem inconsistent to be defining a sustainable city. With more than seven billion people on the planet and a likely maximal population of nine billion or 10 billion, it is not possible to design and build human settlements that are in perfect harmony with nature. According to ICLEI Local Governments for Sustainability (2016), "sustainable cities work towards an environmentally, socially, and economically healthy and resilient habitat for existing populations, without compromising the ability of future generations to experience the same." The goal of the sustainable city is to build human settlements that have the least possible impact on the environment. Although some may question the ethics of this, our definition of *impact* is to ensure that the natural systems central to human well-being are maintained and damaged as little as possible. The sustainable city minimizes its emissions of conventional air pollutants and greenhouse gases; uses as few nonrenewable resources as possible; discharges effluents into waterways after treatment that removes the most harmful pollutants; uses energy and water as efficiently as possible; and attempts to reduce and recycle waste and minimize the impact of whatever waste disposal is needed.

There is no clear, agreed-to definition of a *sustainable city* in the literature, though such definitions often include a range of environmental, economic, social, political, demographic, institutional, and cultural goals (Satterthwaite 1997). In 1991, the United Nations Centre for Human

Settlements (UNCHS) Sustainable Cities Programme defined a sustainable city as one "where achievements in social, economic and physical
development are made to last" (UN-Habitat 2002, 6). A 2013 report by the
United Nations stated that sustainable cities can be achieved when inte-
grating four pillars: social development, economic development, environmental management, and urban governance (United Nations 2013). The
World Bank (2013b) defines sustainable cities as those that are "resilient
cities that are able to adapt to, mitigate, and promote economic, social and
environmental change."

The United Nations Environment Programme (2012, 5) more specifically defines resource-efficient cities as those that "combine greater productivity and innovation with lower costs and reduced environmental
impacts while providing increased opportunities for consumer choices
and sustainable lifestyles." According to Kent Portney, cities that take
sustainability seriously engage in a wide variety of activities that try
to improve and protect the environment, either directly or indirectly
through actions such as reduction of energy consumption. He cites
efforts such as reducing solid waste, redeveloping brownfield sites, protecting biodiversity, improving public transit policy, and enacting climate
action goals as the types of actions that reflect a sustainable-oriented city
(Portney 2013, 122).

It is possible to define *sustainability* so broadly that it loses meaning.
In this work, I will do my best to provide a clear and bounded definition.
In addition to preventing damage to vital ecosystems, the sustainable
city is also a place that attracts people, culture, and commerce. It provides opportunities for human interaction and for activities that develop
human potential. The forms of culture, commerce, entertainment, and
social interaction can vary according to culture, taste, and tradition.
And then the city's function is to provide an opportunity and a facility for these actions to take place. When thinking about a sustainable
city, it is probably worth understanding what an unsustainable city is.
The unsustainable city is one that damages its natural surroundings and
repulses rather than attracts people, culture, and commerce. Cities, like
all human societies, evolve and change. The needs and expectations of
the population change, and a place's ability to accommodate those needs
and expectations also change. And so a city is a set of economic, political,

and social systems that interact with each other and exist in a specific physical setting.

One way to provide a more operational understanding of a city's evolution and of the transition to a sustainable city is to relate my own experience of living in New York City for more than 50 of the past 63 years. When I was a boy, New York apartment buildings still had incinerators in which tenants' garbage was burned in the middle of the night. The rest of the city's garbage was brought to landfills in Brooklyn and Staten Island. Sewage from Manhattan was dumped untreated into the Hudson River. New York was a manufacturing and commercial center. Clothing, toys, bicycles, and even automobiles were made within the five boroughs. The now famous High Line Park was originally an elevated roadbed for freight trains that ran from the Hudson River docks to the factories located on the West Side of Manhattan (in Tribeca and Chelsea). After World War II, nearly half of New York's economy was devoted to clothing manufacturing, distribution, and sales. We had a wonderful, fully functioning system of mass transit, an extensive park system, and a water storage and delivery system that remains an engineering marvel. The water system was needed because we destroyed most of the extensive network of groundwater that lies beneath the street grid in Manhattan and Brooklyn. Toxic waste was stored underground and in waterways such as the still poisoned Gowanus Canal.

Over the past decade, New York City has been gaining population, and it is likely that within the next decade we will be a city of nine million people. That will be a more congested and less pleasant place unless we are able to improve mass transit and build new and more creative public spaces. In the competition for global business and population, a city needs to be safe and orderly, but dynamic and exciting as well. Today, the former factories of SoHo and the West Side either are multimillion-dollar homes and commercial establishments or have been demolished to make way for the shiny new glass and steel creations of the world's "Starchitects." New York City has made the transition from a center of manufacturing, commerce, and finance to a center of education, health care, media, finance, public relations, and tourism. The city now exports all of its garbage to out-of-state incinerators and landfills. Its subway, sewage treatment, water, and park systems provide a tremendous head start in the transition to a sustainable city.

The city's PlaNYC 2030 and OneNYC sustainability plans are efforts to begin the transition to urban sustainability. They set goals and priorities for the public-private partnership that will bring about the transition. The marriage of economic development and environmental protection initiated by former Mayor Bloomberg was both important and innovative. The idea that community-based environmental justice groups and powerful real estate interests could sit together and find common ground was a remarkable accomplishment for the Bloomberg administration and for New York City. In the transition from an industrial and commercial city to a postindustrial global capital, New York City nearly went bankrupt and nearly collapsed in crime and social disorder. But enlightened leadership, resiliency, and luck saved my hometown.

THE RATIONALE FOR A FOCUS ON CITIES

As manufacturing becomes more mechanized, the economies of cities focus on those types of organizations most dependent on people's brainpower and creativity. People are needed less for their muscle than for their brains. These less mechanized and more labor-intensive operations tend to be service providers such as hospitals, educational institutions, hotels, or recreational facilities. Or they can be entities that focus on planning, strategy, creativity, and design—public relations firms, financial advisors, media companies, consulting firms, and cultural institutions.

Globally, more people live in urban areas than in rural areas, with 54 percent of the world's population residing in urban areas as of 2014 (United Nations 2014, 1). The world's population is becoming more urbanized because of an economic change related to the decline of manual labor and the growth of the brain-based economy. While electronic media and communication technology make it possible to contribute creative input from anywhere, the informal network that fuels the creative economy requires that we be physically present to fully participate—something we don't yet fully understand about human communication. The person who "Skypes into" a live meeting is never a full member of the discussion. We are social creatures craving interaction and live contact. This is why we focus on the need to make cities more sustainable: their growth seems related to a long-term change in our economic life and the nature of work itself.

The concentration of population creates some problems for material and energy flows into and out of a human settlement, but it also creates opportunities for economies of scale and creative problem solving. These ideas of closed systems of production and consumption are central to the concept of the sustainable city. As the mechanization of agriculture reduces rural employment and as the Internet communicates the appeal and seductiveness of urban lifestyles, more and more of the world's population is moving to cities. This is especially true of young educated adults: two-thirds of young adults in the United States (ages 25 to 34) with a bachelor's degree live in the nation's 51 largest metropolitan areas (Cortright 2014). This creates opportunities for more efficient production, distribution, and consumption of goods and services. It also creates efficiencies that come from a "sharing" economy. Cars, bicycles, and indoor and outdoor spaces can be more easily shared in a dense settlement. Instead of 200 families each having their own half-acre backyard, a 10-acre park can be shared by many more people and holds the possibility of many uses that require more space than a half-acre.

People move to cities for a range of reasons, from favorable labor market conditions, to attractive public infrastructure, to the benefits of being near centers of finance, corporate headquarters, and information and technology (Buch et al. 2013; Champion 2001; Dittrich-Wesbuer, Föbker, and Osterhage 2008). Cities that have experienced resurgence are usually competitive, attract new and growing activities, and are therefore more interesting places to live; they develop a distinct and comparative advantage (Cheshire 2006; Storper and Manville 2006). Diverse amenities, cultural institutions, educational institutions, and other facilities as well as differentiated neighborhoods are also possible in cities. A neighborhood that attracts families might be distinguished from one that is attractive to single professionals, young couples, and students. The economic and social attractiveness of cities coupled with the diverse character of neighborhoods helps explain the growing importance of urban areas.

On the other side of the equation, the resources required to clothe, feed, house, and stimulate urbanites can strain the resources of the planet if they do not largely depend on renewable rather than finite resources.

Population plus the rate and style of consumption drives resource utilization, but so too does the use of particular materials and sources of energy in production processes. The residents of cities will have some ability as consumers to insist on sustainable production processes, but these are not processes that they will control.

While urban dwellers may not directly observe the environmental impact of their consumption, governments, nongovernmental organizations, and researchers must observe, analyze, project, and communicate those effects. Public awareness of effects will hopefully lead to changes in public policy, regulation, and ultimately private corporate and individual behaviors. Key will be the development and implementation of technologies that permit consumption while mitigating its environmental impact. In the United States, Japan, and Europe, we have already seen that this is possible. Gross domestic product (GDP) has grown over the past half-century, but air pollution and water pollution have been reduced. Control technologies have been put in place to reduce these effects.

Pollution control technologies and green infrastructure cost money but, if designed correctly, can increase quality of life and economic efficiency for people living in cities. When cities are clogged in gridlock, or closed down due to flooding, or waste energy and water, then the cost structure of businesses operating in those cities is impaired because of lower productivity. When air pollution sends children or their parents to the hospital, the costs of health care and child care must be counted as costs of air pollution that can be reduced with investment in pollution control technologies.

In sum, the focus on cities is required because if we are to achieve a sustainable economy and planet, it needs to happen in our cities. The behavior of people and their institutions needs to be changed, and as people will be in cities, we need to focus our attention on these forms of human settlement. The pressure on the countryside and on our ecosystems is coming from the actions of people in cities. As my colleague Ester Fuchs has observed, leadership from government will be required to ensure the focus on cities results in sustainability. According to Fuchs (2012, 53): "Leadership from city government, and especially mayors, is critical to the long-term planning that is required for sustained investment in infrastructure, economic growth and environmental sustainability that will ensure any city's viability in the future."

THE ELEMENTS OF THE SUSTAINABLE CITY

We will discuss in later chapters the social, political, managerial, and economic elements of the sustainable city. The social elements include a set of values and perceptions that lead to consumption and behavioral choices that minimize human impact on the environment. This is facilitated by a legal and regulatory structure that reinforces this behavior. The laws and rules are the outcomes of stakeholder interaction in a political process that is supportive of sustainability. A city's government and private sector must possess the organizational capacity to collect and recycle waste, facilitate distributed generation of renewable energy, build energy efficiency, and ensure the cleanliness of all material flows into and out of the city. And the sustainable city must be capable of obtaining or generating the financial resources needed to develop and maintain sustainability infrastructure (Wang, Hawkins, and Berman 2014).

These elements of infrastructure and rules are essential to the sustainable city and help provide an operational definition of sustainability. The regulatory framework includes the right to be paid for energy contributed to the electrical grid, rules governing waste management from the smallest household to the largest business, building codes, energy efficiency codes, congestion pricing, and other elements of the tax code that reinforce resource efficiency and reuse. The infrastructure includes green solutions to combined sewer overflow, sewage treatment, recycling and effective use of waste materials, water filtration, air pollution control, toxic waste regulation and treatment, mass transit, and electric personal transit.

The most difficult element to build in the sustainable city will be the required infrastructure. This will include microgrids and smart grids that will require a huge investment of capital in rebuilding the electrical system. This will take decades, leadership, and persistence to complete. The same is true of new waste management and recycling facilities and mass transit systems. In the United States, underinvestment in virtually all forms of infrastructure has become normal and accepted practice. Bridges often need to be near collapse before we consider replacing them. In addition to underfunding of capital expenditures, many operating facilities are poorly maintained because of inadequate operations and management budgets. No effort to increase energy efficiency and reduce greenhouse gases can succeed without enhanced mass transit. Because a national solution is not on the political agenda, places like Washington, D.C., New York City, Chicago, and San Francisco are on their own.

New York City's third water tunnel is an example of the nature of these projects. When this water tunnel is completed in 2021, it will have taken nearly half a century and more than $5 billion to complete. The goal of this project is to ensure that the city's upstate water supply can be effectively and efficiently delivered to the city (Flegenheimer 2013). The infrastructure being replaced is close to a century old and badly in need of repair. New York City has a magnificent system of water supply. It is an example of farsighted long-term leadership and investment without which the modern city of New York could never have been built. It takes advantage of ecosystems, gravity, and best-management practices to deliver high-quality and relatively low-cost water to New York. However, like the city's subway system and electrical grid, it is old infrastructure that is decaying, and its maintenance is essential to the transition to a renewable resource–based economy. As a political matter, mayors and other elected leaders prefer capital projects that can be completed within their term in office and are visible and symbolic of progress. A waste management facility, a smart grid, a water tunnel, or a renovated subway line are expensive, sometimes invisible, and difficult for the media to report on.

TRANSITIONING TO A SUSTAINABLE CITY

The job of building a sustainable city atop the current unsustainable city will involve a decades-long transition period and a paradigm shift in the way we manage and pay for cities. The field of management itself will need to change as we integrate the physical dimensions of sustainability into management education and then into organizational management. Just as current CEOs must understand accounting, finance, regulation, international business, strategy, marketing, and human resource and information management, the CEOs of the sustainable city must learn how to integrate energy, water, and material efficiency into routine organizational management along with a concern for environmental effects all the way through the supply chain and the process of production and consumption.

This process has begun in some organizations with the start of sustainability offices, which in some cases are symbolic greenwashing exercises but in other organizations play the role of change agent to remind senior management about sustainability and to provide technical assistance when implementing it. Building the sustainable city requires that more of our organizations have the management and technical capacity to incorporate renewable resources and waste reduction practices into daily organizational life. Environmental risks often become financial risks. The world is too complex and too observed for companies to get away with corporate environmental mismanagement. A company cannot simply dump toxic waste by the side of the road and assume it will not be detected.

My own view is that all competent management should be sustainability management. All of a city's agencies should plan for the effects of climate change on their operations. They should ensure that their buildings and equipment are retrofitted for resiliency in the face of more frequent and intensive storms. Agencies should also make their operations more energy and water efficient, and they should work to minimize the environmental effects of the services they deliver.

In addition to the development of organizational capacity, private and public organizations need to identify means of generating the capital required to construct a sustainable built environment including buildings, energy, and waste and water infrastructure. Finance mechanisms will differ according to the sector and function being performed. Private-sector green finance has become more feasible as investors look for sustainable businesses to invest in. Some funds have been established that require sustainability features in companies being funded. This includes both the service or product being produced and the process of production. Environmental liabilities and costs such as Volkswagen's air pollution issues and British Petroleum's Gulf of Mexico oil spill have been noted by investors, and the risks posed by a lack of attention to sustainability issues have begun to be measured.

The deepest problems will be with the funding of sustainability infrastructure. Because traditional and highly valued infrastructure such as roads and bridges is woefully underfunded, less visible and traditional infrastructure is also suffering from the public's unwillingness to pay the taxes needed to finance these projects. User fees, tolls, and privatization are methods to address these issues, but they too face political opposition. Funding of new green infrastructure such as smart grids, advanced waste management, mass transit, and renewable energy is even more difficult.

One way to assist in financing the sustainable city is to enact laws and public policies that require institutions and individuals to operate according to sustainability principles. Individuals who waste resources or dispose of waste incorrectly can be sanctioned or those who behave "correctly" could be rewarded. Institutions that want to obtain building permits could be required through the building code to build green buildings. Licensed plumbers, architects, electricians, and other craftspeople could be required to be trained in sustainability issues and adhere to sustainability principles. The government could also use its own purchasing power to drive the market toward green production.

For example, New York State Governor Andrew Cuomo took the most significant green step of his governorship when he directed his Department of Public Service to enact a new clean energy standard requiring that by 2030, at least 50 percent of the state's electricity be generated from renewable resources (New York State 2016). This is a demanding but feasible effort and is a clear indicator of the governor's priorities. The renewable energy standard announced by Governor Cuomo is a real, operational, and meaningful step. The New York State Public Service Commission regulates the generation and transmission of electricity in New York State and has a similar responsibility for natural gas, steam, telecommunications, and water. The utilities that the commission regulates are "natural monopolies" due to limits of access to space for power lines and similar infrastructure. When Governor Cuomo directs the Public Service Commission to switch to renewable energy, the force of that order should not be underestimated. It is a meaningful, real-world step that will have a dramatic impact on power generation in New York over the next decade and a half.

Driving local sustainability policy will require active and effective political support at the community level. This political support will need to be built on a strong foundation of popular support for these policies and practices. A necessary but not sufficient condition for the transition to a sustainable city is a demand from the public for the development of such a city. This cannot be a "top-down" approach, else the set of mass behaviors needed in the sustainable city will not take place. Cheating, cutting corners, and disregarding sustainability principles cannot be condoned: this would undermine the legitimacy of sustainability policy and impede the transition to the sustainable city. To build support, sustainability must be defined as a set of positive aspirations, rather than a set of grim behaviors reinforced by negative sanctions. Chapter 3 on sustainable lifestyles will focus on sustainability's positive attributes. These could serve as the basis for political support.

My hope is that all of this will result in more concern for the well-being of our neighbors and our community. That concern could lead to a more determined effort to develop new revenues to pay for infrastructure and to care for each other. It could lead to an examination of our tax structure and an effort to increase taxes where taxation will do the least harm and reduce taxation where it will do the most for society. I know this sounds naive, but my first moment of political awareness was watching a young president ask all of us what we could do for our country, not what we could do for our stock portfolio. If we are to effectively make the transition to a sustainable, renewable resource–based economy, we will need to rebuild that sense of community and shared sacrifice that we have drawn on in the past.

2

SUSTAINABLE URBAN SYSTEMS

———

Defined and Explained

The sustainable city must be built on the foundations of a set of sustainable urban systems. These are production and consumption processes along with infrastructure that enable human settlements to survive and thrive with the least possible impact on natural systems. The operation of these systems requires technology, money, organizational capacity, and political support to be developed and maintained. This chapter will discuss those systems in the contemporary city and describe the changes that will be required in the transition to sustainability.

This chapter defines and explains the following urban systems and defines sustainability in each distinct system: energy, water, solid waste, sewage treatment, food, open spaces and parks, and transportation. First, I will discuss what a sustainable energy system looks like, one that is based on renewable energy and smart grids and that reduces our dependence on fossil fuels. I will then discuss water, a vital resource that must be purified and distributed to everyone in the city and has become a politically sensitive topic. I will discuss how the sustainable city should pursue a number of policies in solid waste to encourage waste reduction, proper waste treatment, and removal. I will then briefly discuss sewage treatment and the need for sewage to be mined for nutrients and chemicals that could be used for other purposes. I will look at some of the issues related to food supply in the sustainable city—how does a city feed its citizens sustainably? I then turn to a discussion of the importance of parks and public open spaces in cities, where people live with less personal space than they might have in rural or suburban places. And finally, I will look at the transportation system and discuss the needs of mass transit and an infrastructure system that are based on renewable energy. This chapter is meant to be a broad overview of the different urban systems that I consider vital for the sustainable city.

ENERGY

All energy sources are subject to federal and state regulations, such as emerging federal carbon dioxide emissions rules, and regulated public energy utilities that deliver electricity to homes and businesses. Yet it is important to think about energy within the framework of the sustainable city. "Cities consume 75% of the world's natural resources, 80% of the global energy supply and produce approximately 75% of the global carbon emissions" (UNEP 2012, 1). According to Portney (2013, 89), "without a doubt, the single most important element in any city's sustainability effort revolves around the environment, and by extension, energy usage and conservation." The energy system includes the electrical grid, as well as delivery of natural gas by pipeline and truck delivery of petroleum to homes for heating and to gas stations for motor vehicles. A sustainable energy system would be based on renewable resources. Most renewable energy is delivered through the electrical grid, but some renewable energy can be generated through geothermal installations, household solar water heaters, rooftop solar arrays, and a variety of other technologies.

A sustainable energy system would involve ensuring access to modern energy services, improving energy efficiency, and also increasing the share of renewable energy in the global energy mix (Ki-Moon 2011, 2). The National Science Foundation (2009, 9) states that "a sustainable energy economy values environmental and ecosystem stewardship, as well as clean, equitable, reliable, renewable, safe, secure, and economically viable energy strategies and solutions." Renewable energy in the United States accounted for about 10 percent of total U.S. energy consumption and about 13 percent of electricity generation in 2015 (U.S. EIA 2016). In 2014 globally, installed renewable electricity capacity represented 28.5 percent of total capacity (U.S. DOE 2014, 4).

A sustainable energy system would also be far more efficient than our current system. Today, a great deal of the energy on our grid is lost in transmission or not put to use during the late evenings and early mornings. It is also inefficiently controlled and used. Electricity transmission and distribution losses average 6 percent annually in the United States (U.S. EIA 2016). This issue is particularly severe in developing countries where losses during transmission and distribution are even higher (Jo and Martin 2015). Poor insulation, inefficient appliances and building systems, as well as sloppy human behavior all waste enormous amounts of energy. The current electrical grid lacks computer controls and in many cases the ability to accept and utilize distributed, decentralized sources of energy. A sustainable energy system would address all of these issues.

At the heart of a sustainable energy system would be a smart grid capable of storing, transmitting, and receiving energy with maximum efficiency. It would also include backup energy systems to prevent blackouts when the grid is malfunctioning. Building the smart grid will be an incremental process similar to the development of the current electrical grid. In all likelihood, the building block of the smart grid of the future will be a set of microgrids built by institutions and communities to provide energy resiliency and backup, to enhance energy efficiency, and to enable buildings and facilities to generate and transmit excess energy. Microgrids are defined as a group of interconnected loads and distributed energy resources with clearly defined electrical boundaries that act as a single entity and can connect or disconnect from the grid (NYSERDA 2016b). Microgrid networks are usually composed of one or more distributed generation sites. These microgrids will have computer controls that will ultimately be knitted together into larger and larger smart grids, made possible by communication technology and computer processing.

Depending on location, distributed renewable energy will include resources such as arrays of solar cells, onshore and offshore wind turbines, and geothermal sources of heat and cooling. Over time, innovation will make these technologies more efficient and less costly. Solar cells are relatively inefficient in their use of solar energy. Some believe that the application of nanotechnology to solar cells will result in smaller, more efficient, and less expensive solar arrays. Energy storage, typically in the form of batteries, is a key technology for sustainable energy. Batteries are coming down in price and size and are becoming more effective and practical. "In multiple application areas around the world, batteries have been deployed to aid the integration of renewable energy, especially solar and wind power. . . . Costs are coming down, and technological progress is improving performance. Recent progress is also making batteries safer and more efficient" (IRENA 2015, 1). The battery market now uses more lithium-ion batteries, which have better cost and performance than other batteries. Battery storage can increase the flexibility of the electricity system, especially in the face of growing renewable energy. An auto battery that delivers 500 miles of driving in one charge could transform the motor vehicle business. A low-priced home battery that allows solar or wind energy to be stored could transform the energy business.

While some of the technologies for renewable energy already exist, the truly transformative technology that would drive fossil fuels from the marketplace has still not been invented. The transition to a renewable energy economy will take several decades to complete. Existing technology is already being used to reduce greenhouse gas emissions while continuing economic growth. Energy efficiency is playing a key role in the early stages of the transition to a sustainable urban economy. But nearly all aspects of modern life require energy. Urban lifestyles require mobility, climate control, food and waste processing, transport, and 24-7 information and communications. Every element of this lifestyle involves the use of technology that requires energy. Our addiction to and dependence on energy makes it very difficult to transition to new forms of energy unless they are as reliable, convenient, and inexpensive as current sources. This is why new technology is needed. If existing technology was truly transformative, the transition to renewable energy would have already taken place.

WATER

While water quality is defined at the federal level in the United States through the Clean Water Act and the Safe Drinking Water Act, potable water for drinking, cleaning, cooking, and bathing must be supplied by the sustainable city. It is a function of local government. This requires water storage, filtration, and distribution systems often costing billions of dollars to construct and many million dollars each year to operate and maintain. "Urban water infrastructure typically includes water collection and storage facilities at source sites, water transport via aqueducts (canals, tunnels, and/or pipelines) from source sites to water treatment facilities; water treatment, storage, and distribution systems; wastewater collection (sewage) systems and treatment; and urban drainage works" (Loucks and van Beek 2005, 427). In addition to the public supply system, private water tanks, pumps, and pipes must be maintained by property owners to ensure that clean water remains clean when it comes out of the faucet. The World Health Organization estimates that 663 million people around the world lack access to improved sources of drinking water, and by 2025 half the world's population will be living in water-stressed areas (WHO 2015). In 2015 and 2016, we saw a crisis develop in the water supply of Flint, Michigan, where unsafe levels of lead were discovered in homes after the city switched water sources. The decision to disconnect a water system from a proven and relatively clean source of water to an unproven source turned out to be a poorly thought through and foolish attempt to save money. This incident raised awareness of the relationship of water infrastructure to effective governance, and media stories in the months following revealed lead concerns in cities all across the United States (Milman and Glenza 2016; Stuart 2016). One USA Today article reported that "an analysis of U.S. Environmental Protection Agency data showed about 350 schools and day-care centers failed lead tests a total of about 470 times from 2012 through 2015" (Ungar 2016). In New York City, the issue of water supply was so politically sensitive that when the New York Times reported a cutback in capital funds to complete the city's vital third water tunnel, the mayor reversed the decision the next day. His staff tried to claim it was a "clerical error." The mayor blamed his staff for poorly communicating his decision. Regardless of the actual events here, the fact is that water supply has become a major political issue in the United States.

There is a sense that like air, water should be free and available to all. It is a necessity for human life, and I think that many believe that one has the right to safe water. While I agree that the provision of safe drinking water is a public responsibility, on a crowded planet it can no longer be obtained free of charge. More and more of our groundwater has been contaminated by normal land uses that involve toxic substances in every-day life. This can include everything from the use of toxics in cleaning fluids, to chemical lawn treatments, to the residue on a suburban driveway after an oil change.

Our lives are built around an almost casual use of plastics and toxic chemicals that have useful properties but may eventually degrade or be damaged by water or fire and then released into the environment. This pollutes the air, land, and water, and while much of it is diluted before humans ingest it, it can enter into drinking water sources that once were clean and can contaminate them enough to require filtration and other forms of treatment. This treatment could take place at the household level if the contaminants are known and stable. Because this is not typically the case, large-scale water filtration and treatment with professional operation and water testing is a more cost-effective approach. This in turn requires that water systems be managed as a public utility, and because of economies of scale, these systems are more cost-effective in urban areas than they are in rural ones.

A sustainable water system is simply one that provides safe and convenient water to everyone. Sustainable water use means that there are "adequate supplies of fresh clean water for present and future generations and for the environment" (Planning Institute of Australia 2006). This requires that contaminants in the water supply be understood, measured, and treated to ensure that the water is fit for human consumption. An unsustainable water system is one that is missing one key component or depends on finite, uncertain, or nonrenewable water resources. Some groundwater sources may be geologic, and some may be recharging slower than the rate of extraction.

In a modern water system filtration is critical, and these processes can use a great deal of energy and may be expensive. Desalinization is becoming more common in some island or arid nations. Recycling wastewater is also becoming more common. While these technologies continue to advance, the technology of sustainable water is more advanced than that of sustainable energy. In the case of water, current technology can be relied on to ensure sustainability; this is not yet the case for energy.

The politics of infrastructure investment will affect water sustainability because elected officials know that they will never get to cut the ribbon on any of these investments, as design and construction can take decades. As I noted, in 2016 we saw an example of the politics of New York City's water supply. On April 5, 2016, *New York Times* reporter Jim Dwyer wrote an excellent, well-sourced, and somewhat depressing article about defunding of the last stages of New York City's third water tunnel by Mayor Bill de Blasio's administration. The third tunnel is needed in order to gradually close the other two older tunnels for repair, ensuring the city's water supply. According to Dwyer's initial story:

> The entire Brooklyn-Queens leg of the new tunnel was scheduled to be finished by 2021, with $336 million included in the capital budget in 2013 by Mr. de Blasio's predecessor, Mayor Michael R. Bloomberg, for whom completion of the third tunnel was the most urgent and expensive undertaking of his tenure. But last year, Mr. de Blasio's administration, eager to keep a lid on water and sewer rates that had grown by an average of 8 percent annually under Mr. Bloomberg, moved financing for the third tunnel to other projects, Amy Spitalnick, a de Blasio spokeswoman, said. The city intends to finish the remaining portions of the tunnel sometime in the 2020s, but it has not set a date for completion nor allocated money in the budget to carry out the work. (Dwyer 2016b)

The negative reaction to the *New York Times* story by the city's movers and shakers, and by many environmentalists, was swift and overwhelming. The drinking water disaster in Flint, Michigan, in 2015 had increased the attention paid to water supply. Moreover, the city had already invested billions of dollars over many decades to build this tunnel. The project was nearly completed; why stop it now? To many people outside of city hall, Mayor de Blasio's decision made no sense. According to the mayor, the newspaper story and the views of his staff quoted in the story were in error. The day after the first story, Dwyer filed a second about the restoration of capital funds to the city's budget:

> Mayor Bill de Blasio said on Wednesday that he was going to add $305 million to New York City's capital budget to speed up work on Water Tunnel No. 3 so that it would be able to serve Brooklyn and Queens. . . . The mayor's announcement came just hours after the *New York Times* reported that his administration last year had removed all money to pay for the tunnel and had also replaced the announced 2021 deadline for

completion with a commissioner's "guess" that it would be ready for service sometime in the mid-2020s. . . . The simplest part of the mayor's day
may have been finding money to pay for the tunnel, not an especially difficult task in a budget swollen with revenues from a booming city economy.
Far more awkward was the struggle by him and his aides to argue that
they had never flagged in their support for the tunnel project, and to
avoid an unflattering comparison to Mr. de Blasio's predecessor, Michael
R. Bloomberg, who drove progress on the construction after work on the
tunnel had moved sluggishly for decades. (Dwyer 2016a)

Despite de Blasio's confused leadership and effort to evade accountability, the real lesson of this water controversy was to reinforce the growing
visibility and importance of the water supply issue.

High-quality infrastructure can often be expensive but must be seen
as an investment in the future. The problem for political decision makers is that reelection is often more important to them than some abstract
notion of "the future." But the future can happen quickly and without
warning. And water resources are not *optional* for a functioning city—
they are necessary. All over America we see older cities with crumbling
infrastructure in need of reinvestment, and in the newer cities of the
Southwest, population growth and anti-tax zealotry put pressure on the
newer infrastructure that was not built to handle the loads it is confronted
with. Many dams in the Western United States are inefficient, and they
lose hundreds of billions of gallons of water each year to evaporation and
leakage underground. Some states are developing plans for new dams
and river diversions, yet as reported in the *New York Times*, "the projects,
coupled with perhaps the most severe water shortages the region has ever
seen, have reignited a debate about whether 20th-century solutions can
address the challenges of a 21st-century drought, with a growing chorus of
prominent former officials saying the plans fly in the face of a new climate
reality" (Lustgarten 2016).

Water is critical infrastructure. Unsafe drinking water can make one ill,
and if children ingest lead, it can cause brain damage. Water is a biological
necessity, and because the primary function of government is to ensure
the security and well-being of the population, protecting a jurisdiction's
water supply can be as important as police and fire services. In New York

City's case, when Mayor Michael Bloomberg took office and asked for an assessment of the risks and threats that the city might face, he learned that a collapse in the water supply system was at or near the top of the list.

Many Americans take their water supply for granted. We turn on the faucet, and clean water flows out. In parts of the developing world, people walk with buckets for miles to find water and bring it back to their homes. The technology of water supply has advanced dramatically in recent decades. Comprehensive water models now use engineering, economic, ecological, hydrological, institutional, and political information to better manage water resources (Loucks 2008). We can also now filter almost any water and make it safe to use. A poor water supply is a function of underinvestment in infrastructure that must be built when the land-use development process degrades traditional sources of water. At one time your home's backyard might have been a place where you could dig a well and obtain clean water. As land development occurs, that clean source can become polluted.

In New York City, there once had been a reservoir at 42nd Street and Fifth Avenue at the site of the present Main Branch of the New York Public Library. As groundwater got polluted and as real estate values soared, the city's leaders realized that they needed to spend the money to go many miles north of the city to store water and to pipe it in. There is always the temptation to do what Flint, Michigan, did and look for a cheaper source of water, but you get what you pay for. As the planet becomes more crowded and as the global trend toward urbanization continues, investment in water treatment and supply needs to grow. For the moment, an awareness of this need seems to have political currency in the United States.

SOLID WASTE

Solid waste, or garbage, removal is a fundamental requirement of the sustainable city. Garbage occupies a great deal of space, can be unsightly, and its open storage can attract vermin and generate disease and illness. As population and consumption have grown, solid waste volume has increased, as have the technologies of waste transport, storage, disposal, treatment, and reuse. Municipal governments must deal with rising costs of waste disposal and environmental effects. The United States consumed 46 percent more materials on a per capita basis in 2000 compared to 1975 (U.S. EPA 2015, 2). The global cost of dealing with all that trash is rising, from $205 billion a year in 2010 to an expected $375 billion by 2025, with the sharpest cost increases in developing countries (World Bank 2013a). In the twentieth century, many cities developed "sanitary landfills," which were essentially either holes in the ground in which garbage was dumped or areas such as wetlands where solid waste was added in order to create new land. Many parts of New York City are land-filled areas. All of Manhattan south of Wall Street is landfill, some of which consists of garbage. For many years, New York City and many other coastal cities barged their garbage out into the ocean and dumped it there. Fortunately, disposal of waste to landfills has decreased: in 2013, Americans sent 53 percent of the waste they generated to landfills, compared to 94 percent in 1960 (U.S. EPA 2015, 5).

Before the twentieth century, most garbage was organic in content and decomposed over time. Modern waste includes many plastics and other substances that are either toxic, non-biodegradable, or both. Such materials and the growing volume of consumption caused solid waste to become an area of urban service delivery requiring technological innovation and increased organizational capacity. Garbage trucks with built-in compactors, landfills, waste-to-energy plants, waste sorting and recycling facilities, and anaerobic digesters (technologies that transform food waste to fertilizer) have all been developed to manage and make use of the growing volume of solid waste. Some of the policies that would be needed for a truly circular economy would require national policy, and in the United States many local initiatives require state approval. There are therefore limits to the discretion that local governments have when setting waste management policies. Nevertheless, there are a number of policies and programs that the sustainable city should pursue regarding solid waste management. The first is waste reduction. Efforts to reduce packaging, including plastic bag fees, are one approach. Deposits on bottles, tires, batteries, and other recyclable items can also be used. Next is building the organizational and technological capacity needed to separate waste into dry and wet garbage (food and nonfood) as well as into more easily recycled waste such as paper and glass. In 2013, America generated about 254 million tons of trash and recycled and composted roughly 34.3 percent of that amount (U.S. EPA 2015, 11). Additionally, we need the organizational and technological capacity to treat waste. This may include waste-to-energy plants and other facilities to either dispose of waste safely or transform it into a usable product. Some of the waste that is burned for energy leaves behind a material that can be used in construction. The U.S. Environmental Protection Agency defines an integrated waste management system as one that has the following four components, in order of emphasis: source reduction and reuse; recycling/composting; energy recovery; and treatment and disposal (U.S. EPA, 2015, 13). All of this suggests an approach more sophisticated than collecting garbage and dumping it into a hole in the ground.

There is also a social element to waste management, which is related to the community's culture and values. One of the goals of a sustainable city is to effectively manage material flows into and out of the city. Garbage, or what environmental engineers call solid waste, presents some of the most difficult challenges to urban sustainability. San Francisco may well be on the way to achieving their goal of "zero waste," or to divert all of its garbage away from landfills by 2020. In 2012, San Francisco claimed that it diverted 80 percent of its waste away from landfills (City and County of San Francisco 2012). At that time, *New York Times* reporter Matt Richtel observed that "San Francisco also has a world-class reputation for its composting processes, which turns food waste into fine, coffee-like grounds that is sent to farms as fertilizer" (Richtel 2016). According to the San Francisco Environment Department, about half of the waste placed in non-recyclable waste bins in 2016 could be recycled, which would drive the waste diversion rate to 90 percent. San Francisco's zero-waste policy includes three goals: prevent waste; recycle and compost; and safely handle toxics. Some assessments of the 80 percent diversion rate state that this number is so high because it includes heavy construction materials and biosolids (MacBride 2013). Nevertheless, San Francisco's unique political and social culture must be seen as a major factor contributing to this program's success. People in that city behave as if reducing waste and recycling are important social behaviors. In contrast, New York City's waste diversion rate in 2015 was 16 percent (City of New York 2016a, 119).

New York City's goal is to get to zero waste by 2030. According to the 2015 Mayor's Management Report, the actual diversion rate was between 1.2 and 2.9 percent below target (City of New York 2015).

Any casual look at New York City's public recycling bins reveals a sense of the difficult road New York must travel to reach anything approaching zero waste. Bins designed to collect paper are filled with bottles, and the bottle bins are filled with a wide variety of unsorted waste. Northern Californians may be thoughtful about waste disposal, but New Yorkers can't be bothered. It is not clear that New York is capable of a cultural shift deep enough to achieve the diversion rates already reached in San Francisco.

Each city is different, and New York's pace, diversity, and size make comparisons to San Francisco difficult. Still, large-scale behavior changes can be achieved with leadership, strategy, and creativity. New York City has eliminated indoor smoking in public places—a goal once seen as unattainable. In any case, behavior change alone is not sufficient. The recycled waste must actually be reused—a problem with the weak market for some recycled substances. The technology of waste sorting and the energy efficiency and cost-effectiveness of recycling also need improvement. A city's system of recycling and waste management is as important as an individual's waste disposal behavior. The technology and market for zero waste will eventually come to New York City, but probably not by 2030.

Zero waste is an element of the concept of a circular economy. In a circular economy, all waste from consumption becomes an input into new production. Inevitably, there is some leakage in the tightest circular production process. But the goal is to move from a linear model of production-consumption-waste to one more closely resembling a circular model. I think of zero waste not as an achievable operational goal, but rather as a model and an aspiration. It is a way to think about resource use and waste management, rather than an absolute target.

It requires a paradigm shift or a new way of thinking about consumption and garbage. Instead of mindlessly tossing something you have consumed into a waste bin, you sort it or consider how it might be reused. In the case of production processes, it includes the concept of producer responsibility. The producer includes in the price of a good an incentive to return the good back to the place of purchase or to the manufacturer.

The manufacturer, in turn, designs the good to be mined for resources or to be reconfigured for additional use. Of course, moving the good or material through the production process requires energy, and so the closed system works best when it is powered by renewable energy.

It is clear to most experts that a system of recycling facilities, waste-to-energy plants, and changed public behavior would be a more cost-effective and environmentally beneficial waste management system for New York City. Unfortunately, because of "not in my backyard" (NIMBY) politics, New York can barely site marine waste transfer facilities to ship garbage away by barge and has been unable to build waste-to-energy plants or other elements of a more advanced waste management system. New Yorkers simply won't accept construction of those facilities in their neighborhoods. In any case, NIMBY politics may well be supplanted by economics. As New York's land prices rise, it becomes increasingly uneconomical to locate large-scale waste facilities within the city.

A sustainable waste management system is one where a city's social norms, organizational capacity, and waste infrastructure operate in synch to reduce the volume and environmental impact of solid waste. It also seeks to recycle as much waste as possible and minimize the use of fossil fuels in operating the city's waste management system. A nonsustainable waste management system simply collects the garbage and dumps it into a hole in the ground. The goal is to think of waste as a resource rather than as garbage to remove and hide.

SEWAGE TREATMENT

Both excess rainwater and wastewater from homes and places of business must be removed from cities safely and with minimal environmental impact. This requires that sewage systems and sewage treatment systems be funded, designed, built, and maintained. The foundation for wastewater discharge control was established through the Clean Water Act, and municipal and industrial wastewater pollutants are limited under the National Pollutant Discharge Elimination System (U.S. EPA 2004, 27). Municipal sewer systems were first built in the United States in the 1850s, and by 1909 most cities with populations over 300,000 were served by sewers (Tarr 1984, 237). The technology of sewage treatment has advanced dramatically over the past half-century. Originally, a sewage treatment plant was simply a screen that filtered some coarse solids from being discharged directly into bodies of water (also known as primary treatment). Now, wastewater treatment is generally categorized into three levels: primary, secondary, and tertiary. Primary, or mechanical, treatment is designed to "remove gross, suspended and floating solids from raw sewage" (World Bank 2016). Secondary treatment removes nearly 90 percent of organic matter using biological treatment processes. These technologies include activated sludge processes and trickling filters. Tertiary treatment can remove 99 percent of impurities through further chemical treatments and is important for public health; water is often disinfected using chlorine or ozone and ultraviolet radiation (U.S. EPA 2014; World Bank 2016). As technology develops, I expect that sewage sludge will eventually be mined for nutrients and chemicals that could be used in food production. A closed system of production and consumption would take food waste and human waste and use them in growing plant life and as a feedstock for animals. In addition to technology that ensures that these waste removal processes do not spread disease, it is important that the energy required to transform waste into product not be generated from fossil fuels.

Sewage systems and treatment are very capital-intensive forms of infrastructure. Pipes and even tunnels must be built to carry the sewage, and in systems that combine household waste with stormwater, a place must be built to hold excess stormwater during intense storm events. Cities are experimenting with green infrastructure, which involves "practices that mimic natural systems into developed areas to manage rainwater where it falls" (U.S. EPA 2014). By adding trees and other vegetation to city buildings, pavement, and other impermeable surfaces, communities can protect their water quality while also providing other social and economic benefits, such as improved public health and reduced urban heat island effects. Permeable surfaces allow water to be absorbed directly by the ground and can also be visual or recreational amenities. Green roofs, porous pavement, rain gardens, and vegetable swales are components that can replace or supplement "gray" infrastructure such as large holding tanks and underground pipes. Natural processes of soils and vegetation can better manage stormwater by capturing, slowing, and filtering runoff.

The key is to avoid overwhelming the capacity of treatment plants to filter and clean sewage before discharging it into the water. The city of Philadelphia, Pennsylvania, has an extensive plan to reduce stormwater pollution through the use of green infrastructure. The Green City, Clean Waters program will re-create the city's landscapes over the next several decades to incorporate green surfaces, and the city expects to reduce stormwater pollution by 85 percent, saving a total of $5.6 billion (Philadelphia Water 2016).

Sewage in the sustainable city must be transformed from a pollutant into a resource. This requires investment in the entire wastewater treatment, transport, and disposal system. We know from human space travel that wastewater can be recycled for reuse. The trick will be to reduce the cost and energy intensity of this transformation. Hopefully, the technological progress now under way will continue.

FOOD

At the start of the twentieth century, about 40 percent of all Americans worked in agriculture, and today that number is a little more than 1 percent (BLS 2015a; Dimitri, Effland, and Conklin 2005, 2). Most of us are far removed from the production of food, and what once occupied most of the time and effort of most people is now the occupation of only a few of us. Without the industrial production and distribution of food, today's cities could not exist. Mass urbanization is in many respects built on industrial agriculture.

There are a number of ways to consider the food system and sustainability. The first issue is sufficiency. Are enough calories per capita available? Is it distributed in such a way that poor people and people on fixed incomes are adequately fed? The Food and Agriculture Organization of the United Nations states that "food security exists when all people, at all times, have physical and economic access to sufficient safe and nutritious food that meets their dietary needs and food preferences for an active and healthy life" (FAO 2008). The second issue relates to the energy intensity of the food. Some foods require a huge input of energy to cultivate and transport to market. Locally sourced food is far less energy intensive. Unfortunately, the areas surrounding the world's megacities do not grow enough food to feed those cities. (Consider that 40 percent of the world's land surface is used for agriculture-related purposes [Walsh 2013]). Despite the popularity of locally sourced food in New York City, only 5 percent of that city's food supply comes from nearby farms. Part of the issue is cultural: people prefer foods that cannot always be grown locally, and they are willing to pay for it. Another part is the sheer volume of need. The New York metropolitan region is home to 20 million people. There are simply not enough farms nearby to feed everyone who lives there.

However, locally sourced produce and urban vegetable gardens can teach city dwellers about agriculture and food production and can increase understanding of the need to protect farm land and to pay attention to the foods that make up our daily diet. Local food options are becoming more popular: the number of farmer's markets operating in the United States grew 180 percent between 2006 and 2014, while the number of schools that house farm-to-school programs grew 430 percent during those years (USDA 2015). According to the National League of Cities (2013), a sustainable food system is one that

- ensures that all residents have access to healthy, affordable food options;
- minimizes the environmental impact of food production and transport;
- facilitates and encourages local food production and processing;
- creates local jobs that provide fair working conditions and a living wage;
- benefits local economies by supporting local food producers, retailers, and businesses;
- maximizes resources through collection and reuse of organics (compost) and other food-related by-products (i.e., fats, oils, grease).

Another key food supply issue is the environmental impact of industrial farming. The mythology of farming as a small, family-owned business has enabled many agribusinesses to be exempted from elements of environmental regulation. Feedlots for livestock have a massive impact on local waterways and ecosystems. Pesticides and fertilizer that mix with rainwater and then run off of farm fields can also impair environmental quality. Many volumes have been written about the environmental impact of industrial agriculture (Beketov et al. 2013; Bouwman et al. 2013; Ponisio and Kremen 2016).

Some types of food production damage ecosystems more than others do. Beef and other meats are probably the most destructive: the Food and Agriculture Organization estimates that livestock production was responsible for 18 percent of human-caused greenhouse gases (Walsh 2013). But many cultures and diets are organized around meat, in part as a status symbol. A reduction in the use of meat in those cultures will come slowly, if ever. A more effective strategy is to develop production methods that seek to minimize environmental impacts. Approaches such as "sustainable intensification" try to increase food production from existing farmland while placing less pressure on the environment—and there are varying technologies and strategies to do this (Garnett et al. 2013). Unlike the other elements of the urban sustainability infrastructure, much of the food infrastructure and many of its environmental effects are geographically removed from the city. That does not make the impact less important, just more difficult to address politically.

The consumption of food, or nutrition and diet, is also an important part of the sustainable city. We have learned that certain foods, particularly in excess, can contribute to obesity, diabetes, and other diseases. One issue related to health and food supply is the existence of "food deserts," or geographic areas that lack fresh fruit, vegetables, and other healthy foods and that are in low-income areas (American Nutrition Association 2015). Scientific studies have suggested that food deserts may negatively affect health outcomes (CDC 2012). Because most urbanites are far removed from food production processes, the regulation of food content and clear labeling of that content are critical elements of public health. Nutrition and control of diet in cities is facilitated by government regulation that requires those selling prepared foods to estimate and visibly post calories, salt content, and other information for customers. There are also health implications of government subsidies. A recent study published in the *Journal of the American Medical Association* found that higher consumption of calories from subsidized food commodities (those related to corn, soybeans, wheat, etc.) was associated with a greater probability of some cardiometabolic risks (Siegel et al. 2016).

Food supply must be seen as a critical part of the sustainable city because urban dwellers are completely dependent on the global food system for their survival. Any break in the supply chain would be catastrophic. According to Kevin Morgan of Cardiff University, urban food security is a relatively new theme but has become a mainstream political issue in the Global North due to unprecedented levels of hunger that followed the 2008 recession (Morgan 2015, 1383). There is a very effective free market in the food industry. There is a great deal of competition, and if consumers are willing to pay the price, virtually any food can be supplied any place at any time. The issue is often price, and that is always related to supply and demand. Although food can be preserved, fresh fish, meat, and produce have a finite shelf life, and that fact also contributes to the impact of market forces on the food system.

OPEN SPACES AND PARKS

In dense urban areas, where many people do not have access to private outdoor space, shared open spaces and parks become important pieces of infrastructure. Some argue that we must *see* nature in the city, as more than mere background for human action, in order to have a functioning model of a healthy city (Capek 2010, 220). There are a great many different types, uses, and benefits of parks. One use is for recreation—ball fields, tennis and basketball courts, pools, skating rinks, swimming, boating, and sailing. In urban areas especially, people visit parks to experience nature and reduce stress. Another use is ecological. Green space absorbs heat and carbon dioxide, reduces air pollution, assists in controlling stormwater runoff, and can help preserve biodiversity. There is also the visual amenity offered by a park. In many cities, homes with a view of a park are more highly valued than identical homes without a park view (Tajima 2003). In PlaNYC 2030, New York City's original 2007 long-term urban sustainability plan, the Bloomberg administration set a goal that every city resident would live within a 10-minute walk of a city park (City of New York 2007). This was a clear, operational, and measurable indication of the importance of parks to urban life. "Urban nature fulfills many social functions and psychological needs of citizens, which make urban nature a valuable municipal resource, and a key ingredient for city sustainability" (Chiesura 2003, 136).

Parks can also provide a commerce-free zone for families. Most public spaces in America feature commercial venues of one sort or another. This adds to the financial pressure on a family. Parks are often free of commerce or, if there is a restaurant or food vendor in the park, it does not dominate the environment. Parks are also a place where friends and families can gather and where neighbors can informally and casually interact with neighbors. They are also a democratizing feature of urban life. There is no VIP line, charge, or special place for the elite in the typical public park. Rich and poor share the same space and facility. In this sense, they can contribute to social understanding and political stability. Increasing evidence shows that natural areas, close to where people live and work, contribute to quality of life. They provide environmental and ecological services, social and psychological benefits such as stress reduction and improved mental health, as well as physical health benefits (Chiesura 2003, 130; Conway 2000; Dahmann et al. 2010; Lee and Maheswaran 2010; Ulrich 1981).

Park design can vary by topography, climate, culture, public demand, and capital investment. Demand for new or redesigned park space will relate to the capacities provided by existing parks. Some parks are focused on active recreation, while others are simply quiet places to sit outside. Parks are an essential piece of urban infrastructure that permit high-density living without diminishing quality of life. In rural areas, people have a great deal of outdoor space that they own. Most of the time there are very few humans in those spaces, but they provide an outlet for people to enjoy the natural world. There is little social engagement in these rural open spaces.

While most people spend most of their time indoors in cities, parks provide actual and implicit outdoor space. The fact is that if everyone indoors suddenly decided to go to their local park, it would be so crowded that it would serve no purpose. But people visit parks for relatively short periods of time, and most people do not visit their local park on any given day. This means that a relatively small amount of land can meet the outdoor space needs of a relatively large amount of people. Skilled landscape design can enable a large number of people to use outdoor park space without being aware of the number of people present. Sound effects from waterfalls can mask the sounds of people. Trees, hills, ponds, and other design features as well as public plazas surrounded by wooded areas without facilities can be used to concentrate people but also leave natural areas less trafficked.

In addition to publicly owned and operated parks, we see examples like New York's Central Park, which is owned by the city of New York but operated by the nonprofit Central Park Conservancy under contract to the city. The Central Park Conservancy raises 75 percent of the park's $65 million annual budget and is responsible for keeping the park maintained and beautiful. It is also possible for private developers to build and operate public spaces for public use or to build and operate private spaces for the use of their own customers or residents. In some cities, a private developer may be given permission to build more densely than the rules allow, in return for the "community benefit" of open space or a public plaza or facility. In some cities, institutions such as museums, botanical gardens, universities, and zoos maintain both public spaces and private spaces that are sometimes open to the public for events. These public-private partnerships can help extend the reach of public spaces.

The Biophilic Cities Network, which was formed in 2013, is a group of leading cities around the world that are pushing for rich, nature-filled experiences in daily urban life. This network is based on the concept of biophilia: "that humans have co-evolved with the natural world, and that we have an innate need for contact with nature and other forms of life" (Biophilic Cities 2015). The network aims to foster urban connections to nature, helping to make cities and urban neighborhoods more resilient and livable through knowledge sharing. To this end, many cities are working to repurpose space for public use, such as previously industrial or contaminated sites, largely for environmental and economic benefit (Pearsall 2010). For example, Fresh Kills Park on Staten Island in New York City is formerly the world's largest landfill. Many cities have success turning brownfields, or "properties that may have hazardous substances, pollutants, or contaminants present," into green spaces (U.S. EPA 2016). In Toronto, Canada, a large brownfield redevelopment process led to soil quality improvement, habitat creation, increased recreational opportunities, and revitalization of many of the city's neighborhoods (De Sousa 2003).

TRANSPORTATION

A fundamental need of the sustainable city is density of population and enterprise. Cities are synergistic and benefit from economies of scale and ease of social, economic, and intellectual interaction. While some of those synergies and some interaction can be accomplished without physical presence, what makes cities special is that they are able to facilitate live, in-person, human interaction. All the urban systems introduced earlier in this chapter are needed for cities to work. But a key element needed to facilitate human interaction is transportation. People need to be able to travel from one place to the other to be with other people and to perform functions that require a particular geographic setting. Universities have meeting spaces, classrooms, laboratories, public spaces, libraries, and other specialized facilities needed to undertake their fundamental research and educational tasks. Hospitals require equipment, laboratories, treatment rooms, systems of quality control, and supplies to deliver health care. Good restaurants need chefs, kitchens, wait staff, and predictable food supplies. You get the idea. People need to get to places to produce and consume products and services that require human interaction.

The movement of people and supplies requires a system of transport. Unmanaged urban density can lead to congestion. The land-use patterns of many American cities, perhaps best epitomized by Los Angeles, relies on personal transportation and sprawls into the countryside, reducing density but increasing congestion due to the absence of sufficient mass transit. Transportation systems heavily influence land-use development patterns. The areas around highway exits become more developed than surrounding areas, but density is sacrificed because of the need for parking. The areas around train stations also become more developed than surrounding areas, but in that case high density can be achieved because parking needs are lower.

As higher levels of density are achieved, the amount of street-level congestion continues to grow. This makes subway construction and service more essential and more cost-effective as well. However, it is far more cost-effective to build subways *before* high levels of population density are achieved. The price of land and the cost of disrupting existing development can be quite high. Moreover, if mass transit is built before high levels of development occur, the transit can be used to direct development in ways that can minimize environmental effects and the cost of infrastructure. According to Portney (2013, 117), "cities that take sustainability seriously try to integrate transportation planning with other types of planning, including residential planning and zoning, industrial and job site location, and other issues." And according to the American Public Transportation Association: "Increased public transportation investment can lead to significant economic growth, as a consequence of both the short-term stimulus impact of public transportation outlays and a longer-term, cumulative impact on economic productivity" (Weisbrod, Cutler, and Duncan 2014, iii).

For people of limited means, mass transit holds the potential of being a method of lowering the costs of entering the mainstream economy. There is no capital required to purchase a vehicle and no need to obtain a license to drive one. A recent immigrant can ride alongside the long-established citizen from home to work or school. Like parks, this collective resource can have the effect of democratizing the society. To the extent that such access decreases extreme poverty and provides a sense of hope and opportunity, it can have a positive impact on political stability.

In the United States, transportation accounts for 70 percent of petroleum consumption and 27 percent of greenhouse gas emissions (U.S. DOE 2016). As society moves toward a renewable resource–based economy, electric trains and buses purchased by governments and that run on renewable energy can help accelerate the transition from fossil fuels. Differential pricing of mass transit can enable governments to accomplish a number of social goals. Schoolchildren can be given free transportation. The elderly can receive discounted fares, and fares can be reduced during off-peak hours to help reduce congestion. People of limited income can be given discounts to encourage them to participate in work or learning. Personal transit is more difficult to influence, although tolls and congestion pricing can be used to change traffic patterns and subsidize mass transit. An electric car, powered by renewable energy could allow autos to be less environmentally destructive.

Even in places with good mass transit, personal transportation will always have an important role to play in the sustainable city's system of transport. In some locations, and for some purposes, personal transport is the best option. But the use of private vehicles can be reduced with ride-shares, taxis, and bicycles, which can be used alongside private autos. In some cities, different means of transport are used to connect together. A park-and-ride lot can allow people who are not within walking distance of mass transit to connect with a subway. Bike-share facilities and ride-share networks can also be used to bring people to and from mass transit resources. Public bike sharing can be found in more than 600 cities in 52 countries around the world (Yin, Qian, and Singhapakdi 2016). There are about 104 cities in the United States that currently operate bike-share systems, and 77 percent of bike-share docking stations can be found near local public transportation stops, such as bus and commuter rail stops (Firestine 2016). A combination of market forces, publicly funded infrastructure, subsidies, and user fees can be used to develop a complex public-private transit system that facilitates mobility while reducing environmental impact. Such a system is a key piece of infrastructure for the sustainable city.

CONCLUSION: THE URBAN SYSTEMS REQUIRED FOR THE SUSTAINABLE CITY

Most of us live in cities, and many of us who do not still live urban lifestyles—connected to the places that generate the ideas, entertainment, culture, social norms, technologies, economic trends, and political regimes that influence all of us. While most of us live apart from nature, all of us depend on nature and on ecosystem services that produce our food, water, and air quality. The human-made systems described in this chapter can be designed to minimize environmental effects and contribute to the sustainability of natural systems or they can damage and even destroy natural systems. The goal of the sustainable city is to use the concentration of people and resources to generate the capital and financial resources needed to build and maintain systems that minimize environmental damage.

It might be possible to build human-made systems that could replace natural systems. We do not know how to do that yet, but I have no doubt that someday we will possess that knowledge. My hope is that before we develop that capacity, the value and ethic of environmental protection will be held so deeply that preserving the planet for its own sake will be politically feasible. In the present, we do not have the technology to supplant nature and we are not yet good enough at leaving the planet, so we have no choice but to attempt to build human settlements that preserve nature. This chapter briefly summarized the systems that cities are built on and provided an overview of how those systems might be designed to minimize environmental effects. I now place an additional lens on urban sustainability and in the next chapter will explore the concept and practice of a sustainable urban lifestyle. In that chapter I ask: What do people do in the sustainable city? How do they spend their time? What is the nature of their work, their recreation, and the other elements of their lifestyle?

3

THE SUSTAINABLE LIFESTYLE

Defined and Explained

I t is clear that sustainable urban systems lead to a sustainable environment and economy only if they provide support to what we might call a sustainable lifestyle. This is a way of life that sees material consumption as a means rather than as an end and attempts to ensure that the materials consumed have as little negative impact on the biosphere as possible. Definitions of *sustainable living* in the literature generally refer to using as few resources as possible, reducing carbon footprints, and reducing environmental damage (Regenerative Leadership Institute 2007; Winter 2007). The United Nations Environment Programme (2011, 6) defines sustainable lifestyles as "rethinking our ways of living, how we buy and what we consume but, it is not only that. It also means rethinking how we organize our daily life, altering the way we socialize, exchange, share, educate and build identities."

We live in a global mixed economy that is largely market based but also involves a great deal of public-private partnerships, government regulation, government services, and transfer payments. Capitalism is organized around giving consumers what they want. But those wants are far from fixed and vary by demographics, place, culture, and history. In this chapter's thoughts about the sustainable lifestyle, I will begin by discussing the economic changes now under way in both consumption and the nature of work. I will then turn to a discussion of social change and its impact on sustainability, particularly the trend toward reurbanization and urbanization. Then I will discuss education and sustainability. As we move from a manual labor–based economy to the brain-based economy, the need for constant education and training is part of the overall notion of a sustainable lifestyle. I will conclude with an analysis of sustainable lifestyles in the sustainable city.

THE ECONOMIC CONTEXT OF A SUSTAINABLE LIFESTYLE

Environmental advocates often focus on individual behavior and say we need to develop lifestyles that consume less and do not damage ecosystems. On a worldwide basis with billions of people aspiring to higher levels of material consumption, individual reductions in consumption in the developed world will have little real impact. But I have hope that we can and are changing the nature of consumption just as we are changing the nature of work. Economic consumption is based on exchanging accumulated wealth for products and services. The prices charged in the exchange are not related to the size of the material involved in consumption. The biggest car need not be the most expensive one. If you buy an expensive software application, view a movie, or buy a subscription to an online game or streaming music service, the only material used in the exchange is energy—which someday will be renewable. You are paying for the creativity and intellectual power of the people who create the application, movie, game, or music.

We spend more and more of our time each day interacting with our social and professional network via cell phone, text, or e-mail and absorbing information we receive from the Web. Annual expenditures for cellular phone services increased from $608 per consumer unit in 2007 to $963 in 2014—an increase of 58.4 percent (Creech 2016). This represents a form of consumption. The business models that generate revenue for the organizations that create and disseminate information and entertainment content vary, but they all generate revenue that represents payment for work. Sometimes content is paid for by item, sometimes by monthly subscription, and sometimes in exchange for watching an advertisement. Increasingly, the higher value-added parts of our economy are not in the production and distribution of material goods, but in the design, creation, and marketing of both products and services. And this is a trend that is accelerating. The hardware in our laptop is now advancing at a much slower pace than the software. In fact, it will be the need for additional computing power, speed, and media capacity coming from new apps that will drive the development of new hardware.

The twenty-first century brain-based economy has changed the nature of production and consumption. The majority of today's generation consists of millennials, generally defined as those born between 1983 and 2000, currently between the ages of 16 and 33 years (US PIRG Education Group 2014). (Other definitions differ slightly: the Pew Research Center defines millennials as those born after 1980, while researchers Howe and Strauss define the group as those born between 1982 and 2004 [Bump 2014; Pew Research Center 2016]). We spend our time differently today than we did in the twentieth century. We spend money on different things—a recent Morgan Stanley report indicated that young people are spending more on rent, cell phones, and personal services and less on apparel (Lutz 2015).

Peer-to-peer markets, known as collaborative consumption, or more commonly "the sharing economy," also demonstrate changes in the way we consume and use goods and services. The sharing economy has its origins with the founding of eBay and Craigslist in the mid-1990s; both are online marketplaces for recirculation of goods (Schor 2014). Zipcar, an American car-sharing service where individuals pay a subscription and then rent cars for short periods of time, began in 2000. Perhaps the best example is Airbnb, a travel lodging, Web-based community marketplace for people to list and book housing accommodations around the world. Airbnb offers unique travel experiences, as individuals are able to stay in another person's home, apartment, or villa (even castle) instead of a hotel. Airbnb has served more than 60 million guests since it was founded in 2008 and is now in more than 34,000 cities in 191 countries (Zervas, Proserpio, and Byers 2016). Another example is the growth of ride-sharing services such as Uber and Lyft, which offer transportation that is accessible through mobile applications. An individual needs only an account and the smartphone app to submit a trip request, which is then routed to Uber drivers who use their own cars. Uber is in 482 cities worldwide while Lyft operates in nearly 215 cities in the United States and 9 cities internationally. Young people in particular are the target market for ride-hailing services: a 2014 survey from the University of California, Berkeley found that 73 percent of such rides were taken by 15- to 34-year-olds (Rayle et al. 2014).

We also work at different jobs. Projections of the next decade indicate that our aging population will require more health-care workers and that manufacturing will continue to decline in the United States. According to the Bureau of Labor Statistics:

> Service-providing sectors are projected to capture 94.6 percent of all the jobs added between 2014 and 2024. Of these 9.3 million new service sector jobs, 3.8 million will be added to the healthcare and social assistance major sector. . . . Healthcare and social assistance is projected to increase its employment share from 12.0 percent in 2014 to 13.6 percent in 2024. Construction is projected to add 790,400 jobs by 2024. Even with these additional jobs, employment in the construction major sector is not projected to return to the 2006 peak. Manufacturing employment, between 2014 and 2024, is projected to decline at a 0.7 percent rate annually, a more moderate decline than the 1.6 percent rate experienced in the prior decade. (BLS 2015b)

Where we work and what we do is also dependent on the job market and status of the economy. A report by the Office of the New York City Comptroller analyzed the impact of the 2007–2009 Great Recession on jobs, specifically for individuals entering adulthood. The unemployment rate in New York City after the recession was more than 18 percent for those between 18 and 29 years of age, compared to 10 percent for the overall national unemployment rate (Office of the NYC Comptroller 2016). And even though the unemployment rate has since declined, millennials employed in New York City in 2014 earned 20 percent less than their counterparts in previous generations (millennials here are defined as those born between 1985 and 1996).

The nature of work has also changed with the growth of the "gig economy." Research indicates that the proportion of American workers who don't have traditional jobs (who instead work as independent contractors or freelancers) has soared in the past decade. The number of Americans using these alternate work arrangements rose 9.4 million from 2005 to 2015, which is greater than the overall rise in employment. The percentage of workers in these arrangements was 15.8 percent in fall 2015, up from 10.1 percent a decade earlier. Construction and professional and business services were the two most prevalent industry groups among those employed in alternative work arrangements in 1995 and 2005, but the educational and health services industry has surpassed them over the past decade. By 2015, more than one in five workers in an alternative work arrangement was working in education or health services (Katz and Krueger 2016).

These changes relate to the global economy that the United States participates in and in many respects continues to lead. While manufacturing output will grow in the twenty-first century, much of it will be largely automated. This has already happened to agriculture, and the trend continues. Our labor needs are changing, and the resources required to buy material goods are also in relative decline. This means that the proportion of our income devoted to food and clothing and maybe someday to shelter is a declining portion of our income and therefore of our overall consumption. There is less need for unskilled manual labor and more need for skilled, service-oriented labor. More and more of the labor involved in manufacturing requires the skilled operation of complex machines. Event planners, software designers, communications strategists, policy analysts, Web designers, personal trainers, health-care workers, social service providers, and countless other service professions are replacing the "butcher, the baker, and the candlestick maker."

The evidence of this change is everywhere. You can see it in the physical changes on the West Side of Manhattan near my home. The best example of this in New York is the now world famous High Line Park. The High Line was originally an elevated roadbed for freight trains that carried raw materials and manufactured goods back and forth between the West Side docks and nearby Manhattan factories. The docks are gone, replaced by a beautiful park and a growing number of amenities; the manufacturing firms are gone, replaced by Google, media companies, fancy restaurants, and even fancier apartments. The elevated train track is now a beautiful, world class, path-breaking public park. Technology forced the change. The West Side docks could not accommodate containerized shipping and so the freight port moved across the river to New Jersey. The factories themselves sat on real estate that was too valuable for the old functions of packing meat, making clothing, and fabricating other materials central to the economy of the first half of the twentieth century.

The changing economy challenges our labor force and our educational system, and in the mid-1970s it almost drove New York City into bankruptcy. But New York City struggled and then managed to adapt and change. In the process of changing the economic role of the city, we need to pay more attention to the impact of our production and consumption on the environment and on all elements of the supply chain that bring goods and services to us. Building of systems that reduce environmental effects is more important than individual consumption patterns. That is not an argument for people deliberately wasting energy and other resources, but rather that to have a real impact, we need to embed new patterns of consumption in more sustainable supply chains. Individual behavior is not enough. We need institutional behaviors that ensure production and consumption with the least possible impact.

Tangibly, that means our food waste and sewage need to be reprocessed for fertilizer. The waste stream must be mined for resources and, to a growing extent, reused for new production materials. It goes without saying that all of this will require increasing amounts of energy, but that energy must not be generated by fossil fuels. There are other system-level changes needed, particularly our educational system, which must do a better job of preparing people for the professions of the future rather than the jobs of the past.

And what of sustainable consumption? It will happen to the extent that our consumption shifts from filling our closets with shoes to filling our minds with ideas and our souls with friendship, relationships, music, film, theater, culture, and dance. Visits to the mall may be replaced by visits to cafés to argue politics with friends or watch the ball game, or by trips to the gym, the ball field, the basketball or tennis court, or by a hike in the park. And much of this move toward sustainable consumption is driven by young people. In the United States, millennials account for more than 70 million people with more than $200 billion in purchasing power. "Their generational impact on politics, economics, social and cultural constructs, etc., are similar in magnitude to that of the Boomer generation, but magnified exponentially by the power of the Internet and its related technologies and applications" (Bigi, Corvi, and Ng 2007). And according to a survey by Elite Daily and Millennial Branding, many millennials are using their buying power to support specific social and environmental agendas (Schawbel 2015).

It is not difficult to imagine these changes, but the only way they will happen is if people are positively attracted to them rather than punished for their attraction to nonsustainable consumption patterns. According to researchers from the University of Groningen, by creating a dynamic in which pro-environmental behavior is not only the "right" thing to do but also aligns with the "norm" of society, those behaviors become what is referred to as "normative goal framing." Observing others participating in a sustainable behavior can encourage one to adopt those habits as well (Steg, Lindenberg, and Keizer 2015). Researchers who have examined interventions to increase environmentally friendly behaviors found that the key to success is linking those behavior changes to shared values (Miller and Prentince 2016). Culture and values are far more powerful forces of social change and consumption patterns than regulation. Prohibition didn't end drinking. If someone wants to buy 50 pairs of shoes and ride around in the water on their speedboat that should be their right. But hopefully the images of interesting and exciting work and play will reflect the growing understanding of the need to minimize the damage of our work and play on the planet that sustains us.

CONSUMPTION AND THE SUSTAINABLE LIFESTYLE

This shift in consumer attitudes demonstrates that mass behaviors that lead to a sustainable and renewable economy will not come from a simple desire to protect the environment. Hybrids and electric cars are more expensive than traditional autos, and as long as that is the case, they will have difficulty competing with them. Price, quality, reliability, and convenience all factor into consumer decisions, and no one should be surprised that lower gasoline prices change the consumer's calculus when he or she buys a big car.

Sustainable consumption is playing a key role in the movement toward a sustainable economy (Yin, Qian, and Singhapakdi 2016). A *Time* magazine article reported in 2009 that: "We are again entering a period of social change as Americans are recalibrating our sense of what it means to be a citizen, not just through voting or volunteering but also through commerce: by what we buy. There is a new dimension to civic duty that is growing in America—it's the idea that we can serve not only by spending time in our communities and classrooms but by spending more responsibly" (Stengel 2009).

Few people are in love with the internal combustion engine, and a cheap, convenient, and reliable electric car would win in the marketplace. But the movement away from hybrids to larger cars provides an excellent example of the limits of "responsible consumerism." Like corporate social responsibility, these notions only work when they line up with self-interest. Enlightened self-interest makes sense. Consumer and corporate behavior is by definition about self-interest. Corporations are organized to make money. Consumers spend their money on goods and services that provide maximum benefit. We want their behavior to be "responsible," and it is good when they get there on their own, but it makes little sense to build a system around altruism.

The challenge of sustainability is the seductiveness and appeal of our current mode of consumption and our way of life: instant information at the push of a button; instant entertainment at the push of the same button; creature comforts like climate control; plentiful, fresh, and tasty food; mobility; and time for leisure and relationships. The list is nearly endless. No one is going to willingly give that up, and billions of people on the planet are aspiring to share in that lifestyle. We need to transform economic production and consumption while increasing its volume. A higher proportion of economic consumption will include less of material-based goods and services and more of education, research, entertainment, social interaction, crafts, art, and physical fitness.

The transition to a sustainable economy requires that we transform our production and consumption processes to reduce their impact on natural planetary systems. This means that young people must learn about those systems, how they function, how we make use of them, how we endanger them, and how we might protect them. Our goal should be to maintain our way of life and expand it to others who seek it, without damaging the planet.

THE SOCIAL CONTEXT OF A SUSTAINABLE LIFESTYLE

None of us is capable of predicting the future, but the optimism of young people is even more impressive to me when we think of the uncertainties of the world to come. These include

- the sustainability challenges of climate change and ecological damage from economic development;
- the changing nature of work and the impact of technology on work, families, communities, people, and the planet;
- the unpredictable impact of a world economy, global media, and culture;
- increased sectarianism and tribalism in seeming response to the force of the global megaculture and the seductiveness of modern lifestyles; and
- the ever-increasing technology of destruction, particularly when coupled with evil and terrorism.

The nature of work is changing, and the opportunity structure is a moving target that is difficult to understand and advance within. Change can be scary but seems inevitable. I am comforted by the successful transition of my home city, New York, from industrial-era disintegration to today's postindustrial world capital. It was not a smooth transition, but it worked. I am reassured by the optimism of young people as measured by pollsters. The transition to a sustainable economy will require an optimistic human spirit, ingenuity, and creativity. The path and skills needed to get there are still being defined, but perhaps out of this uncertainty our children will find their place in the world and build a world that has places for them to find.

The world we are creating with global communications and global economic production is fast becoming a true global society. Nevertheless, place—or home—continues to have meaning as both communities and nations become even more treasured in a world of mass-produced impressions and fear of homogenization. Today, young people are building a shared understanding of how the world works through social and other Web-based media. Millennials in particular turn to social media for an authentic look at what's going on in the world, especially content written by their peers whom they trust (Schawbel 2015). The balance of local and global must still be determined. There is little point arguing about the desirability of this new world. It is happening and I see no prospect of it stopping. The trajectory of the new global society and the impact of these media and the new ones that will follow are difficult to predict.

According to the Centers for Disease Control and Prevention: "The trend in U.S. life expectancy since 1900 has been one of gradual improvement. In 2010 life expectancy at birth was 78.7 years, an increase of 11% since 1970" (CDC 2013). In 1900, Americans, on average, lived to be 47. Today we live over three decades longer. Science and technology have helped make our lives longer and better. There are many reasons that we live longer today than in 1970, but a cleaner environment is one of them. While you could argue that with enough improved medical technology, exercise, and food and diet, we could tolerate the costs of more polluted air; who wants to be part of that experiment? Who wants to allow their children to be part of that experiment? And why do it? There is no reason to trade off economic development and environmental protection—we can and should have both.

The demand for sustainability is both the cause and effect of a number of facts of modern life, particularly growing population, increased urbanization, increased use of natural resources, pollution, climate change, the political demand for economic development, inexpensive information and communication, and the growth of a connected global economic and communication system. These forces pretty much define the twenty-first century and the world we now live in.

We need to see increased acceptance of values and conscience related to global sustainability, and I see far more of that today than I have ever seen. Along with the growth of economic consumption, we have seen the growth of an environmental ethic. Importantly, the environmental ethic is stronger in young people than in old people. Young people have grown up hearing about environmental issues, along with issues related to nutrition, physical fitness, and public health. Between 1981 and 2014, the number of Americans who were dues-paying members of private health clubs (gyms) quadrupled from about 13 million to more than 50 million (Cortright 2015b). People are paying more attention to their own well-being and to the state of the planet. Many believe that their elders have destroyed the ecosystems that they will end up inheriting, and that perception has created a sense of urgency about the sustainability of our planet and the global economy that we have built.

Young people are bringing sustainability to the center of the policy agenda and to the center of the management agenda in the organizations they work for. In the workplace, it is not unusual to see staff asking for "greener" workplaces—toxic-free workplaces that are energy and water efficient and that pay attention to the environmental effects of the goods and services they produce. In a survey of millennial consumers, 75 percent said that it's either fairly or very important that a company gives back to society instead of just making a profit (Schawbel 2015). Large companies such as Apple and Walmart have been greening their operations in response to both external market pressures and internal social demands within their organizations. In addition, according to a Bolster Creative workplace trend report, millennials want and expect more from a job; they want a purpose, meaningful work-life experience, diversity, and sustainable products. Six in 10 millennials say that part of the reason for working for their current employer is a due to sense of purpose that the company provides (Bolster Creative 2015).

We are all more sensitive to human impact on the environment than we were a generation or two ago. According to a 2014 Nielsen Global Survey, 55 percent of consumers worldwide are willing to pay more for goods and services from socially responsible companies. And many people are more aware of the impact of the environment on their own health and well-being than they once were. People pay more attention to food, exercise, and health care. Parents extend their deep concern about their own exposure to toxics to an even more profound concern about the impact of environmental degradation on their children's health and future. The investment world is even seeing a difference: 71 percent of individual investors are interested in sustainable investing, and millennial investors are nearly twice as likely to invest in companies or funds that target social or environmental outcomes (Morgan Stanley Institute for Sustainable Investing 2015).

While issues like climate change and biodiversity may be difficult for some folks to understand, issues like toxic waste, water pollution, and

especially air pollution can assault one's senses and are really easy to
understand. Young people have been seeing, smelling, and even touching
these environmental insults all of their lives. Even if the environment in
the United States is in many ways cleaner than it was in 1970, the country
is more crowded, consumes more, wastes more, and is more aware of the
connection of pollution to health.

What is most interesting to me as I view all of this from the sustainability perspective is that the consumption of information and ideas has
little negative impact on the environment. If we spend more of our time
engaged in education, culture, and entertainment, presumably we spend
less time consuming, and possibly destroying, nature. While there is some
danger that the joy of engaging in the natural world might be forgotten,
I am too much of an optimist to believe that will happen. As more of us
live in cities, our exposure to wilderness, and even nature, will continue
to decline, but perhaps our appreciation for nature need not suffer.

We are already more removed from the natural world than America
was in 1900 when 40 percent of us worked in agriculture. But we continue
to add acreage to our parks, and here in New York City, OneNYC, our
update of the PlaNYC 2030 sustainability blue print, set a goal that all
New Yorkers live within a 10-minute walk of a park. President Obama
preserved more land and bodies of water than any president in American
history. Real estate values reflect the economic value of trees, water, and
sunlight. And of course, we still require the food, water, and material
comforts the planet provides. We do not need to live in nature to know
that we depend on it.

But there is no denying that the forces of modernity are powerful. The
technologies of information, stimulation, and comfort are seductive and
addictive. But so too is the pull of place, family, friendship, loyalty, and
love. In the end, humans are social and emotional creatures, and we crave
company and interaction. More than consumption patterns shape our
values; people are not simply "consumers." These other values are the basis
for a sustainable society supporting a renewable economy.

At the heart of the struggle for sustainability is a human society that is
creative, ingenious, often courageous, and always deeply flawed. Straight-
line projections of food supplies and demand, energy supply and demand,
and even global warming create warnings that we should always heed, but

never despair over. If the past 50 years has taught us anything, it is that one should never project the next 50 years. People and their needs remain constant. Little else seems to be very stable. I guess that's what makes the world interesting and why we are all so drawn to human expression as conveyed by the globally communicated world of endless images, voices, and ideas.

REURBANIZATION AND THE SUSTAINABLE LIFESTYLE

In some parts of the United States, we are starting to see our culture and economic life shifting in a direction that is less destructive of the planet. People are moving back into cities. Over the past decade, American cities in particular have experienced a renaissance—young professionals and their familiars are increasingly choosing to live in the urban core (Katz and Bradley 2013). Their recreation often includes activities that consume few resources: viewing media, creating art, exercise, social engagement, and outdoor activities. In the brain-based economy, an increasing portion of wealth comes from "software" rather than "hardware." But even when we consume material goods we are doing better. We are learning that a well-designed home that is smaller and uses fewer finite resources can be constructed to be quite beautiful and can be both comfortable and more sustainable.

An analysis by City Observatory reported that over the past few years, American cities have grown faster than outlying areas. City centers are more competitive in 2011 than they were in 2007, and there is growing evidence that the health of the city center is related to the health of the overall metropolitan economy (Cortright 2015a). The move to cities is in part fueled by young people. Since 2000, in the nation's largest metropolitan areas, the number of young adults with a 4-year degree living in neighborhoods within 3 miles of a city center has increased 37.3 percent (Cortright 2014). This movement of young educated adults to city centers is providing an important impetus to urban economic development across the country. U.S. Census data indicate that metropolitan areas grew at a faster rate than the rest of the country, especially cities like Austin and Seattle. According to chief economist for the National Association of

Homebuilders, David Crowe: "There's been a surge in urban apartment building. . . . The 25- to 34-year-old age group is focused on living near their peers. They want be socially engaged and live near work. They want to reduce their automobile use. All of those things aim at high-density, urban-type living" (Frizell 2014). And according to researchers School-man and colleagues (2014): "College-educated millennials, motivated by a preference for vibrant, walkable neighborhoods with access to good public transportation, are helping to drive an economic resurgence in many American cities."

A lifestyle of conspicuous consumption and endless shopping can often result in many wasted hours in traffic and be quite frustrating. More and more people are attracted to a lifestyle that allows them to reduce their driving and rely on walking or mass transit for most of their needs. In a survey of millennials by Global Strategy Group (2014), 80 percent reported that it is "very" or "somewhat" important to have a wide range of transportation options, including public transportation and car- and bike-sharing services. More than half (54 percent) of millennials surveyed say they would consider moving to another city if it had more and better options for getting around, and 66 percent say that access to high-quality transportation is one of the top three criteria they would weigh when deciding where to live (Global Strategy Group 2014). Between 2001 and 2009, the average number of miles driven by 16- to 34-year-olds dropped by 23 percent (US PIRG Education Fund 2014). Lifelong education, live music and theater, bars, and other forms of entertainment are more likely to be plentiful in cities, and both young people and old people are gravitating to these places. Research has found that Americans are increasingly seeking "third places" to socialize outside of the work and home; for example, the number of coffee shops in the United States nearly doubled in recent years, from 11,000 in 2003 to 20,000 in 2012 (Cortright 2015a, 21). New York City has about a million college and graduate students. The presence of health care, mass transit, elevators, education, entertainment, restaurants, and culture has caused some observers to term the Upper West Side of Manhattan a "Naturally Occurring Retirement Community" (NORC).

A recent *New York Times* article highlighted the types of cities that are doing well, hiring people, and where consumers are spending. It mentions places like Seattle, Portland, Atlanta, Orlando, and Denver. Patricia Cohen reports: "Such places have become business incubators and magnets for educated millennials. The lifestyles that 20- and 30-somethings often seek depend on a medley of urban living, public transit and lots of entertainment options" (Cohen 2016).

And while an increasing percentage of our population is living in cities, the countryside is not being abandoned. There appear to be plenty of people to populate both rural and urban areas. The fight over fracking in the Northeast is largely a battle over rural development and conservation. The intensity of that conflict can also be seen as an indicator of the continued value placed on nature. It is also one element of the battle between forces of local identity and forces of the global economy and society.

LIFELONG LEARNING: A KEY ELEMENT OF THE SUSTAINABLE LIFESTYLE

We have created a complex world with many moving parts, and we require many forms of what I'll call micro-expertise. We often need someone who is good at setting up spreadsheets and breaking data down to its key elements; someone who knows how to use social media to raise funds for a charitable event; or someone who knows how to set up a home entertainment system and explain its upkeep to a person who is more than 50 years old. What part of our current education system is designed to produce the agile problem-solver and lifelong learner who can figure this stuff out and put it to work?

The knowledge base needed for the renewable economy will require basic understanding of our world but also will require many specialized areas of technical expertise. Technical expertise is an essential factor in resource allocation and in implementing sustainability initiatives overall (Wang, Hawkins, and Berman 2014). The point is that one size will not fit all, and a quantitative test of knowledge, while useful, is only a very partial indicator of learning and knowledge. Understanding a person's ability and knowledge may be the most difficult measurement task there is.

The complex world we live in requires multiple forms of expertise, orientation, and talent. Sustainability problems often require knowledge of chemistry, hydrology, toxicology, ecology, finance, politics, law, management, and marketing. Is anyone good at all of that? One person might be good at running the numbers. A second knows how to conduct interviews that provide the backstory that the numbers don't convey. A third could be a genius at graphic design and does a wonderful job of laying out the slide deck and report. The final product is a test of the group's ability.

In a world where manual labor provides fewer and fewer jobs, knowledge and learning are more important than ever. But we are basing our definition of education and learning on the economy and knowledge base of the nineteenth and twentieth centuries. We live in a different world now. At one time, an educated person was someone who had read the hundred or so most important works of Western civilization. Those books remain vital, but they are not enough.

Education must be redefined to include group work and problem solving, must be lifelong, and must be made more accessible to poor children and older people who need to learn to participate in the nation's economic, political, and social life. The ultimate measure of our educational system will be the wealth, fairness, and sustainability of our system of economic production and consumption. Let's measure that.

New methods of education, such as online courses and programs, provide students with flexibility, cost savings, and increased connections globally. With the widespread availability of high-speed Internet, online learning options (which began in the early 1990s) have become increasingly popular. Distance learning creates a virtual classroom and a new mode of education where students and professionals can learn new topics, gain valuable skills for their jobs, and even complete entire degrees. Massive open online courses (MOOCs) have gained popularity in recent years and enable people across the globe and of all ages to have access to information and training that could give them the tools needed to succeed in the new global economy. First introduced in 2008, MOOCs are online courses aimed at unlimited participation and open access via the Web. A *New York Times* article called 2012 "the year of the MOOC" (Pappano 2012). According to the European Commission's Open Education Europa initiative, as of January 2015 there were more than 3,842 MOOCs worldwide, with a growth rate of 201 percent in 2014 (PRNewswire 2015). Compared to traditional online courses, MOOCs are usually free, do not offer formal academic credit toward a degree, and can have thousands of participants for one class. The basic idea is that free courses can bring education to the most remote corners of the world, help people advance their careers, and perhaps expand intellectual networks.

Data suggest that most of the students taking online courses are not doing it for a degree: a survey of 35,000 MOOC students in 2014 found that 79 to 86 percent of students already have a college degree (LeBar 2014). Many students take these courses to help develop skills applicable to their current job or to help them get a new one. Of course, because many courses are free, and anyone can sign up, some students are poorly prepared for university-level work. Only a few complete the full course, and there have been issues with grading and cheating. However, many argue that completion rates should not measure the success of the MOOC: MOOCs have benefits even if not completed and can be taken to offer new knowledge, supplement other courses, or introduce topics of interest.

Teaching young people to value the planet and the resources it provides is the way to ensure that our potential to solve the sustainability problem is achieved. Sustainability began to be incorporated into education practices around the world in the mid-1990s, formally recognized in the United States in 1996 when the U.S. President's Council on Sustainable Development published "Education for Sustainability: An Agenda for Action." An increasing number of mainstream school systems and institutions are incorporating sustainability paradigms at the K–12 level, with positive effects. This growth is evidenced in changes in school curriculum to include sustainability and environmental concepts, incorporation of sustainability in state standards, increase in research and funding for environmental education, and even "greening" of schools themselves (Federico and Cloud 2009). What strikes me about the growth of sustainability in K–12 education is its capacity to be truly transformative. Evidence supports the relevance of environmental education concepts to other areas of education, with positive student outcomes in reading, math, and science, and other broader skills in critical thinking, leadership, and engagement (Athman and Monroe 2004; Church and Skelton 2010; Ernst 2007). For example, in one survey by the nonprofit Facing the Future of 55 science teachers who had incorporated climate change into their high school classrooms, 96 percent reported that the climate change lessons helped increase their students' overall critical thinking skills (Church and Skelton 2010). "The earlier we lay the foundations of sustainability education for K–12 students, the sooner our higher education colleagues will be able to advance the requisite knowledge, skills, and habits of mind in our young people" (Wheeler and Byrne 2003, 25). My hope is that a generation of young people educated from pre-K on to understand the sustainability challenge will approach it with a fresh and innovative perspective.

This means that sustainability curriculum must teach scientific, economic, social, cultural, and political facts and concepts to help students understand sustainability problems but must be careful about prescribing solutions. Dogma and environmental determinism need to be avoided. Straight-line projections about the future based on current trends should be avoided and replaced by multiple projections based on varied assumptions. History is difficult enough to understand; predicting the future is as much craft as it is science. The challenges that humanity faces can be addressed, but they require a deep understanding of the trade-offs caused by our modern lifestyle. The sustainability curriculum at all levels should stress interdisciplinary exchange and learning, teamwork, and problem-solving methodologies.

Two-thirds of students attending a 4-year bachelor's program leave with an average of $26,600 in debt (American Student Assistance 2013). There are now roughly 43.3 million Americans with $1.23 trillion in student loan debt (Student Loan Hero 2016). Just a decade ago that number totaled $300 billion (Korkki 2014). A survey conducted by American Student Assistance found that those with student debt are delaying decisions to buy a home, get married, have children, save for retirement, and even enter a desired career. In this survey, 63 percent said their debt affected their ability to make larger purchases such as a car; 73 percent said they have put off saving for retirement or other investments; and 75 percent indicated that student loan debt affected their decision or ability to purchase a home (American Student Assistance 2013).

The transition to a renewable economy requires education at every level. We need students in public and private schools to develop a deeper understanding of the global sustainability crisis, but we also need aspiring professionals and current professionals to develop the expertise needed to begin the transformation in real time, today. In organizations all over

America and all over the world, young people are pushing older people to develop and implement sustainable practices and organizations. Many professionals are returning to school to learn about sustainability, and many aspiring professionals are focusing their undergraduate and graduate education on sustainability science, engineering, policy, management, architecture, design, communication, and art.

CONCLUSION: UNDERSTANDING THE SUSTAINABLE LIFESTYLE

What is a lifestyle? It is a weird word, but it is what people do with their time: work, recreation, entertainment, travel, social life, family life, religious life, education/learning, hobbies, and so on. It also includes the setting within which they undertake these activities: where someone lives, where they work, where they hang out, where they pray (if they pray).

What does any of this have to do with sustainability? A person can spend his or her time and enjoy that time by consuming resources at a ferocious rate or at a moderate rate. You could, for example, live on a 5-acre estate with a 15,000 square-foot home and work out in your own private gym, and entertain in a home theater with a 30-seat screening room, swimming pool, and barbeque. Or you could live in a 1,500 square-foot apartment, work out at a commercial gym, go to a movie theater, swim at a public beach, and have a meal in a restaurant. You could walk and bike to work, take a train, or be driven in a huge, shiny SUV. You could recycle your food waste from your kitchen or toss it out your window to the alley below.

A majority of Americans today (85 percent) commute to work in private automobiles. Only 1 in 20 people commute to work via public transit, although this number is different for cities that have excellent transit systems, such as New York and Chicago. In addition, there are some signs that we have passed the period of "peak driving." Vehicle miles traveled per person have declined over the past decade, and the younger generation is getting licenses later and driving less than previous generations of young adults (Davis and Baxandall 2013). Even outside of the United States, the percentage of young people with a driver's license has recently decreased substantially (Sivak and Schoettle 2012).

Your lifestyle has resource implications. Even the huge home could be designed with geothermal climate control, have a solar water-heating system, and be built to minimize runoff from the driveway to the nearby stream. It's not simply what you do, but how your lifestyle affects natural systems.

All of us inevitably consume resources in the course of our daily lives. We plug our computer into the electrical supply, we turn on the climate control, we turn on the lights; we bathe, dress, and eat. How we spend our time is changing. Part of this is the changing nature of work and the fact that work is no longer limited to the office or factory or to particular times of day. In the global economy, the workday is always beginning somewhere. The Internet and cloud computing mean that analytic work and written work can take place anywhere at any time. So too can meetings. They can become Skype sessions or conference phone calls. While I remain convinced that humans require live interaction and in-person contact to be effective, a high proportion of communication is electronic and requires few incremental resources to be undertaken. I am quite certain that we spend more time than ever communicating professionally and personally.

This communication process is part of the knowledge or brain-based economy. More and more of our time is involved in learning and communicating ideas, events, activities, and information. In a sustainable world we can develop our bodies, brains, as well as our relationships with each other, and we can create products, services, culture, art, science, and technology while paying attention to how we affect the planet's basic systems.

This does not require a monolithic, one size fits all, limited way of life. You can build a zero-energy house on the outskirts of Houston and drive your electric car all over. Or you can live in an apartment in Portland and bike, walk, and take the light rail. What unifies the people pursuing a sustainable lifestyle is that consumption is a means and not an end. The winner isn't the one who accumulates the most stuff, but the one who lives the fullest life—however that is defined. For some it may be in service to one's community, for others it might be nurturing a grandchild. For some it may be creating a new way of sharing autos or homes or inventing an application or a new way of managing an electric microgrid.

These choices are made possible by an economy where less and less of the gross domestic product (GDP) is devoted to the manufacturing of food, clothing, and shelter. At one time, that was virtually all the economy did and was how people spent all of their time. Today, we spend less of our time pursuing our basic needs. This means that more of our work and our time must be devoted to other pursuits. To be clear, we cannot survive without food, air, water, clothing, and shelter. But through automation we need fewer people to make those things. The problem is that all of that technology requires energy, and so the quest for new renewable energy technology is critical to our long-term well-being. Fortunately, we are making rapid progress and I believe that soon, that progress will be fully reflected in the marketplace. The next economic transformation will be built on the implementation of a decentralized model of renewable energy generation and distribution. Fossil fuels will be driven from the market-place by this new technology and method of delivery.

With the energy dilemma fixed, we will then be even freer to pursue the postindustrial economy and the sustainable lifestyles we are beginning to see. The transition to this new economy will not be easy, and it is likely that many people who benefited from the old economy will have diffi-culty adjusting to the new one. It will be the job of government to ensure that the social safety net is adjusted to provide not just material well-being, but a sense of purpose and dignity for people who face the chal-lenges of adjustment. The repair and construction of twenty-first century infrastructure could provide the bridge employment needed by people with twentieth century skill sets. While those construction jobs are also increasingly mechanized, our roads, bridges, electrical system, and water and waste systems need a major infusion of capital and construction.

Funding that infrastructure will compete with revenues devoted to transfer payments to our aging population. One way to fund that would be to create incentives for people to work longer into their lives, retire later, and draw down less of those payments. Another would be a national service requirement for young people to devote themselves to civic purposes or military service. Funding for that program and work-ing later in life would require changes to Social Security and Medicare, the third rail of American politics, where few elected leaders are willing to tread. It would require a sophisticated discussion of the transition to the new economy that is well beyond the capacity of our current national political life.

Listening to some contemporary political dialogue in America, it is hard to believe that people will be prepared for this very different global economy. The idea that we can turn back the clock, or would even want to, has political resonance with some. A politics of nostalgia is easy to understand but does not solve the problem of an evolving economy. It is a strategy that cannot succeed because it promises employment in jobs that technology and the market have already discarded.

Our economy and our lifestyles will continue to change as technology and new services and products come to market. At one time, we all sat still and watched a show on television at the same time, because that is how the product of television entertainment was served up. Today, we watch a show whenever we want to. At one time we had to go to the office to access the files, people, and information we needed to get the job done; today, we can access what we need for work from anyplace. How we spend our time and what we do every day will continue to change. Human ingenuity guarantees it. What is not guaranteed is that our inventiveness will take into account the health of our natural systems. But the growing number of people determined to live a sustainable lifestyle will help ensure that this new chapter of economic evolution will not be the final chapter.

4

THE TRANSITION TO SUSTAINABLY MANAGED ORGANIZATIONS

This chapter focuses on organizational management and the growth of organizational interest in the physical dimensions of sustainability. Organizations have begun to factor environmental and resource issues into routine analysis and decision making: they are now regularly looking at energy consumption as a way to cut costs and increase productivity and are taking a closer look at how they use water and other material resources. The costs and impact of waste, discharges of effluents, and emissions are now subject to routine analysis. This is not only for regulatory compliance but also to ascertain the risk of environmental effects and the cost of insuring such risks. This chapter will discuss these issues and provide examples of organizations starting to include the physical dimensions of sustainability as basic management inputs. I will also discuss the trend toward incorporating sustainability into organizational strategy. I will look at how sustainability is reflected in an organization's incentive system, in its approach to financial risk and reporting, and in its marketing as well as training and capacity building.

The concepts of the circular economy and its related engineering field, industrial ecology, are also elements that enable us to move toward closed-system production and consumption. This reflects the development of sustainability as a basic engineering, economic, and management concept. I will discuss the evolution and status of organizational sustainability metrics and reporting. As management scholar Peter Drucker once said, "you can't manage something unless you can measure it." Without measurement, you can't tell if management's decisions are making the situation better or worse. We are a long way from having generally accepted sustainability metrics. But many organizations are developing their own internal measures for the elements of sustainability they want to manage. Finally, I will discuss sustainability *mismanagement* and look at companies whose failure to incorporate sustainability into their organizational structures has literally cost them billions of dollars.

The sustainable city will be built on a foundation of sustainably managed organizations. Organizations are the building blocks of human production and consumption. Our ability to make the transition to a renewable resource–based economy will require that we weave together a broad set of sustainably managed organizations and institutions. While the previous chapter focused on individual sustainability behavior, this one is focused on collective action, or the behavior of organizations.

THE SUSTAINABLE SUPPLY CHAIN

Most twenty-first century organizations are part of interorganizational networks. Unlike the vertically integrated industrial giants of the start of the twentieth century, they need not own all parts of production to deliver products or services. This presents challenges to sustainably managed organizations, because they do not always have information about or control over the production processes of their suppliers. For example, the mining companies that provide the metal inputs for electronic equipment often have no direct link to the electronics companies producing the products. And while most global organizations are well versed in sustainability initiatives, their suppliers can lag significantly behind the curve.

It is not sufficient for organizations to simply implement sustainability initiatives within the walls of the company; it must extend down the supply chain. Twenty-first century managers are quickly learning that supply chains are only as sustainable as their worst participants. In 2007, Mattel had to recall almost one million toys because they contained lead-tainted paint from a subcontracted supplier in China (Story 2007). When Nintendo received a score of zero on its conflict minerals report card in 2012, the company was forced to take a closer look at its supply chain. In a statement to CNN, Nintendo said it "outsources the manufacture and assembly of all Nintendo products to our production partners and therefore is not directly involved in the sourcing of raw materials that are ultimately used in our products" (Sutter 2012). In 2015, Chipotle Mexican Grill's stock price plummeted after an *Escherichia coli* outbreak in its food sickened hundreds of people. Chipotle depends on a complex supply chain for its 1,900 outlets that includes scores of small, independent farmers, and because of this complexity, the company still doesn't know which ingredients made people ill (Berfield 2015).

Most organizations now recognize that a sustainable supply chain is no longer an optional "add-on" but is critical to the success of their operations (Accenture 2016). The business case for ensuring supply chain sustainability is clear. By managing and improving environmental, social, and economic performance throughout supply chains, organizations can conserve resources, optimize processes, and increase productivity. The number of companies dedicating sustainability resources to supply chain efforts grew from 10 percent in 2014 to 26 percent in 2015 (GreenBiz 2016). As more companies focus on sustainability, they are asking producers for help in meeting targets. Since 2010, Walmart has eliminated more than 28 million metric tons of greenhouse gas emissions from its supply chain by implementing various innovative measures across both its global operations and those of its suppliers (Walmart 2015).

However, while more organizations have expanded their sustainability programs to include suppliers, many struggle with implementation. Lack of agreed upon metrics and concrete action make it challenging to drive sustainability performance along supply chains. Sustainably managed organizations must also consider the risks posed to supply chains by climate change and changing weather and water supply patterns. Global organizations depend on supply chains stretching around the world. Extreme weather, floods, and droughts can cause major disruptions and create ripple effects all the way up the supply chain. Sustainability and risk management professionals have a crucial role to play in identifying, quantifying, and mitigating risks from climate hazards.

We need to embed new patterns of consumption in more sustainable supply chains. Individual behavior is not enough. We need institutional behaviors that ensure production and consumption with the least possible impact.

SUSTAINABILITY MANAGEMENT

Sustainability management is simply the latest step in the evolution of the field of organizational management. Twentieth-century managers were concerned with finance, human resources, information, production, performance, marketing, strategy, and globalization. Today's managers must also pay attention to the use and cost of natural resources, the cost of waste production and disposal, and the environmental effects of organizational outputs and waste. These *physical dimensions of sustainability* can no longer be ignored. They are an increasing percentage of an organization's cost structure.

The fundamental concept of sustainability management is that CEOs and COOs must know enough science to manage these physical dimensions of sustainability; specifically, water quality and quantity, toxicity, waste, energy efficiency, environmental effects, and the impact of toxics on ecosystems and human health (Cohen 2011). Just as a manager must be able to read a financial statement and understand an analysis by a marketing focus group, that manager must also understand enough science to make decisions about an organization's use of and impact on natural systems. Just as managers need to manage finance experts, they must be capable of managing technical experts.

Both management education and managers themselves must meet the challenges of complex contemporary conditions. Energy, water, and other raw materials are becoming a larger element of the cost structure of all organizations. Water is no longer free. Waste disposal costs more and more each year. These increased costs are not limited to manufacturing but include the growing number of service organizations as well. Microsoft, Walmart, and thousands of other organizations are learning to build these factors into their operations and their internal control systems because these are sound management practices that lead to the accomplishment of an organization's goals. While some of those goals may be reputational, some are also related to lowering the costs of operations.

If we are going to develop a sustainable economy with a high level of production while maintaining the planet's ecological health, we need to get better at setting and complying with environmental rules. We no longer live on the frontier or in the Wild West. With more than seven billion people on the planet, we need to learn to be more thoughtful about how we produce and consume the material resources we need. Air pollution rules are not optional because breathing is not optional. A recent report from the World Health Organization (WHO) found that more than 80 percent of people living in urban areas that monitor air pollution are exposed to air quality levels that exceed WHO limits (WHO 2016).

A key element of sustainability is the business principle of sustainability management: the organizational practices that promote a renewable economy. Compare it to total quality management (TQM), which is a way to manage organizations to improve their overall effectiveness (Porter and Parker 1993). TQM is a technique used to reduce waste and improve quality in an organization's supply chain and production processes. Sustainability management is a way to reduce resource costs and the potential costs of environmental effects caused by an organization's production and consumption.

The field of organizational management is undergoing a fundamental transition, one of a number that have taken place over the past century or so. At the start of the twentieth century, management needed to understand mass production and the first complex supply chains, and so, under the theory of Taylorism, or scientific management, labor was mistakenly seen as an almost machine-like part of the production process. Eventually, we learned that people were not machines, and management embraced human resource development. After the Great Depression, we developed generally accepted accounting principles (GAAP) as the standard framework for financial accounting, and at that point all CEOs had to learn a little accounting and how to read financial statements. From the 1960s through the 1990s, we saw the price of information and computation start to drop, and subsequently CEOs needed to learn information technology (IT)-based performance measurement and management. At the start of the twenty-first century, developments in communication and transportation stimulated the global economy, and CEOs found themselves managing international enterprises. Global supply chains and interorganizational networks have replaced many vertically organized hierarchies.

Today, as our growing economy damages the ecosystems that make human life and wealth possible, CEOs must manage the physical dimensions of sustainability. Both service and manufacturing organizations must manage their inputs of energy, water, processed and raw materials, their production of waste, and the environmental impact of their activities and actions of their consumers. What used to be a small, negligible part of management has become a core function of management. Today, more than 60 percent of executives consider sustainability "very" or "extremely" important to overall corporate strategy, according to a McKinsey global survey (McKinsey and Company 2010). Companies are recognizing that sustainability has benefits such as cost savings, access to capital, product innovation, reputation, and better employee engagement (Haanaes et al. 2011). While America's business schools have generally not yet figured this out, I believe that by the end of the next decade the definition of competent management will include sustainability management. Competent managers will be sustainability managers.

Through regulation, technology, and better management, we have begun to learn how to build our economy while improving our air, water, and land. While we have begun this learning process, we have a long way to go. People know it is important that natural resources be used carefully and need to be protected from irreversible harm. That does not mean we have the finances, organizational capacity, and technology needed to do the job, but at least we know it is a job we need to do.

The typical management concerns will remain, but in today's more complex and crowded world, CEOs must be even more sophisticated than ever before and pay attention to the physical dimensions of sustainability. The costs of ignoring environmental risks can be quite high. The costs of preventing environmental damage end up being quite low when compared to the costs of remedying environmental disasters.

We need to develop the organizational capacity to understand our planet well enough to manage and control our interactions with it. In a planet growing to as many as 10 billion people by 2050, with global gross domestic product (GDP) rising, we must learn how to manage and maintain a high-throughput economy that does not destroy the planet's ability to sustain life—especially our own. We need to understand what our production and consumption patterns and behaviors do to natural systems. We need to adjust our behavior to maximize production while minimizing damage. We need to do this to ensure that a reasonable facsimile of the lifestyles that many of us live today can be shared across the planet and can be maintained for us as well.

This will not be easy to do, but building this organizational capacity is the main goal of the field of sustainability management. The argument for doing this is we do not have any real choice. The technology of mass destruction requires political stability for human survival. Economic development is required for long-term political stability. People must have an ownership stake in society and a vested interest in its stability. Without that, we simply will not survive.

INCORPORATING SUSTAINABILITY INTO ORGANIZATIONAL STRATEGY

Corporate governance provides lessons that may be useful here. Before the Great Crash of 1929, every corporation's financial reporting was an individual matter. If you invested in a company back then, it was sometimes like gambling at a casino. A company's financial reports might assert that they were making money, but who knew? After the crash, America started to regulate financial markets, and by the end of the 1930s we had developed GAAP. If a company wanted to raise capital through the public marketplace and sell shares through an American stock market, the Securities and Exchange Commission required strict and audited financial reports. Companies complied in order to have access to the public marketplace to raise capital. Companies had to accept the new regulatory standards for financial reporting or they would be denied access to those capital markets. Government was an essential partner in setting up and enforcing rules of the game that corporations benefited from.

It is becoming obvious to many observers that in addition to understanding the possible financial risks posed by corporate performance, we also need to better understand the environmental risks posed by a company's behavior. Environmental risks often become financial risks. The world is too complex and too observed for companies to get away with corporate environmental mismanagement. A company cannot simply dump toxic waste by the side of the road and assume it won't be detected. Every smartphone in the world holds the potential to provide a record of corporate misbehavior. The nongovernmental organization (NGO) that identified Volkswagen's software deception was small, but effective. To manage in this new world, global leaders and CEOs need to understand and practice sustainability management.

Sustainability management makes it possible to build an economy that allows a high standard of living without destroying the planet that sustains us. It requires organizations to think about their use of water, energy, and other raw materials and about the waste resulting from production and consumption. It requires that we use incentives to change behavior, that we promote new thinking about resource use, waste, and impact, and that we integrate sustainability into routine management and decision making. It also requires that we develop "generally accepted sustainability metrics"—a way to measure an organization's or jurisdiction's progress toward sustainability.

This environmental awareness, which could be labeled a paradigm shift, is exerting pressure on many of the day-to-day actions routinely undertaken by corporations, government agencies, and nonprofits, and on behaviors seen in communities and households. Individual behavior is changing as well, as we examined in the past chapter. People think about how long they run the faucet. They think about the bin they toss their garbage in. They think about the cleanliness of the air they breathe. The change in private organizations can be striking, as people assert the need to protect the environment along with typical corporate goals such as profit, market share, and return on equity.

Here in the United States, a large number of cities, corporations, and nonprofit institutions are moving to increase their use of renewable energy and are taking steps to make their buildings and vehicles more energy efficient. For example, in 2014, New York City Mayor Bill de Blasio launched "One City: Built to Last," a 10-year program dedicated to retrofitting public and private New York City buildings to increase energy efficiency. Some of this is driven by an effort to reduce costs and make energy systems less centralized and more resilient. As technologies advance in information, communications, and manufacturing, energy stands out as an old-fashioned, highly centralized, technological backwater, ripe for innovation and cost reduction. We are hearing more boardroom discussion of smart grid technologies, distributed generation of energy, energy efficiency, and innovative models of financing renewable energy systems. Efforts to improve energy efficiency are increasingly common in cities, companies, universities, and hospitals. Planned corporate investment in energy efficiency is at its highest level ever, and this trend seems more than likely to continue in the foreseeable future.

This is not an impossible task. We simply must move past short-term expedience and the type of thinking that argues: "In order to make an omelet you've got to break some eggs." We need to use our analytic, information, and communication resources to do a better job of managing human impact on the environment. While this may raise some costs in the short term, it will lower costs in the long term. As we get better at managing our activities, we will learn more about how to simultaneously produce and protect, and the price of protecting the environment will go down.

Today's corporations, nonprofits, and governments are operating on a more crowded and interconnected planet that provides great opportunities, but also poses great threats. The threats are not simply coming from competitors but from the actions of organizations. Greater care and thought must be devoted to the use of natural resources and the impact of an organization's production, outputs, and consumption on ecosystems. It is not enough to call production errors "breakage" and add them to the cost of doing business. Those are twentieth-century principles of "macho-management" that have no place in the complex environment of the twenty-first century economy. Managing organizations on a planet of three billion people, as we had in the 1960s, is different from managing organizations on a planet of more than seven billion people as we have today. It is not simply a difference of degree but of kind.

In addition to these operational issues, organizations must learn to develop a sophisticated and productive relationship with the regulatory environment. Compliance with environmental regulations, labor law, the tax code, and occupational health and safety rules are part of operating and managing a modern organization. While it is true that many nations do not enforce their environmental and occupational health and safety rules, a quick study of economic history demonstrates that the trend is toward more enforcement rather than less enforcement. And even when the government ignores noncompliance with the law, NGOs and consumers notice it. In 2015, Volkswagen's noncompliant vehicles, with doctored software that faked emission results, lost market value faster than Volkswagen's stock.

One of the key factors in developing and managing an organization's strategy is to set priorities and focus attention on a limited number of factors. Many businesses believe that regulation disrupts the free market and is rarely needed. It would be smarter for them to accept the rules and learn to operate creatively within them. One adapts to the changing business and regulatory environment and learns to thrive under new conditions. The world is changing. Technology has changed it, and the rules and business environment will continue to be a moving target. Managers must get used to this new world and learn to adapt and thrive. Managers have a key role to play in advancing sustainability by developing ways to integrate resource conservation and efficiency in the operations of their organizations and by managing environmental risk. Within the next decade or so, all competent management will be sustainability management, and managers who do not understand the principles of sustainability will be considered incompetent and out of date dinosaurs.

This shift is already under way. There is an increasing trend among companies to incorporate intangibles and sustainability concepts into financial reports. According to research in the *Journal of Applied Corporate Finance,* this new model is a move away from the traditional linear financial statement model that focuses only on product sales (Adams 2015). And there are many examples of companies who are leading the way. In 2010, Unilever launched its Sustainable Living Plan under the leadership of its CEO, Paul Polman, and has since made considerable progress incorporating sustainability into its central business model. Under the umbrella of its comprehensive overall sustainability strategy, Unilever is utilizing its wide array of brands to target distinct social issues, invest in sustainable technologies, and change consumer behavior. Similarly, in a path-breaking move in 2014, outdoor clothing company Patagonia announced the dissolution of its sustainability department with the intention to "integrate innovation sustainability thinking, values, and goals into every employee" (Perella 2016).

When judging the competence of a corporate management team, investors are beginning to examine the organization's capacity to measure and manage its use of resources and its impact on the environment. When people and businesses are looking for places to move to or grow, they look at the adequacy of a jurisdiction's energy supply, water supply, air quality, transportation system, and overall quality of life. In a mobile global economy, people can choose where they live or work.

Within organizations, sustainability should be integrated into an overall strategic planning process. While it may be necessary to begin with a distinct and identifiable sustainability plan, in the long run it should be part of an organization's routine planning and management system. Specifically, this should include the following:

- **Goal setting**: The sustainability goals for the year and for the next 3 to 5 years should be set in clear, measurable terms. This can include improved energy and water efficiency, reduced greenhouse gas emissions, recycling rates, and similar measures.
- **Development and maintenance of organizational capacity**: An organization serious about making the transformation toward sustainability must detail the human and financial resources allocated to achieving sustainability goals. Achievement of sustainability goals must be assigned to specific people and organizational units. These need not be new units or even units labeled "sustainability," but the priority of these goals and the tasks involved must be clear.
- **Sustainability metrics**: It is critical that the achievement of sustainability goals be measurable and that those measures be included in individual performance metrics and organizational incentive systems. Not only is it true that what gets measured gets done, but what does not get measured is ignored and considered unimportant.

Sustainability is dismissed by some as a fad and a form of public relations. It certainly has those dimensions. But in management terms, it can be considered as the logical extension of W. Edwards Deming's work on quality management. Deming focused on driving waste out of organizations. While much of his attention focused on wasted work, wasted supplies were also an element of his concern. An organization that uses materials more efficiently and finds productive uses for its waste will be more competitive than an organization that ignores these costs. We are already seeing an increasing integration of the sustainability perspective in corporate culture, driven by both consumer demand and by real cost savings. According to the GreenBiz "State of the Profession" report, 89 percent of Fortune 500 companies publish some type of sustainability report (GreenBiz 2016). And 79 percent of Fortune 500 companies reporting to the Carbon Disclosure Project (CDP) had higher returns on their carbon reduction investments than from their overall portfolio (CDP 2015).

Moreover, the type of risks incurred by ignoring the environmental effects of an organization's actions are famously illustrated by British Petroleum's oil rig explosion in the Gulf of Mexico, Volkswagen's doctored emissions software, and General Electric's dumping of PCBs in the Hudson River. These environmental mistakes have cost these companies billions of dollars. Avoiding these mistakes entirely is impossible, but reducing the probability of these errors is possible and an indicator of competent management.

INDUSTRIAL ECOLOGY

Industrial ecology is the process of designing production processes and products with the least possible impact on ecological systems. It is a way of thinking about the life cycle of a product or a service that examines the finite resources utilized and the pollution this product or service releases into the environment. A central component of industrial ecology is "to move from a linear to a closed-loop system in all realms of human production and consumption" (Lowe and Evans 1995). It is an attempt at an "industrial mechanism that is consistent with nature's metabolism" (Huber 2000, 269). The goal is to close the cycle of production so that production is based to the extent possible on renewable resources, and whatever is not renewable is recovered.

In that respect, one solution to waste management is non-technological. It involves designing products that can be easily reconditioned and reused and designing a postconsumption process that brings the product back to the manufacturer. Xerox does this by leasing some copiers and designing them for remanufacturing. Hewlett Packard does it by designing its toner cartridges to be easily collected and then refilled.

Just as economic development creates a demand for more energy and exacerbates the climate crisis, increased consumption results in more waste. With growing wealth, we will see growing garbage. We've already seen it in China, which surpassed the United States in 2004 as the world's largest waste generator, and we will see it everywhere before long. But in the United States we have started to put in place local solutions to this very local problem. Europe and Japan have been ahead of us on this issue for more than half a century. Managing our waste, the back end of the economy, is at least as important as managing our consumption.

This points to the need for all of us to understand the interconnectedness of our system of production and consumption and the need to develop system-wide solutions to system-level problems. Take the issue of electronic waste, or the garbage produced when we dispose of outmoded phones, computers, or entertainment systems. Consumers alone cannot address the issue of electronic waste. The companies that manufacture and sell these devices must play a key role. While most global companies incorporate sustainability strategies into their manufacturing and logistics processes, only a fraction of these companies have programs to lower the environmental impact associated with the consumer's use of their products. I won't get into the short product life cycle of iPhones or their planned obsolescence, but if these products are going to be replaced with such great frequency, then Apple and its competitors must design the phones to be easily disassembled for raw materials or for components to be used in other devices. This effort to recycle should become a design parameter for engineers working on the next generation of electronic toys. In addition, companies should be offering bounties for consumers to trade in their old model to receive a discount on the new one. The production chain from producer to market should become a closed loop, with the product going back to the manufacturer or to an organization capable of making some use of it.

I suspect that the pattern we have recently seen with personal computers may become more prevalent with smartphones and other technologies. PCs have become commodities that have few distinguishing features. The exciting new consumer uses will increasingly be seen in software changes rather than in new hardware. That may have the effect of reducing the rapid increase in electronic waste volume. Of course, the growing market for electronics in the developing world will offset some of that environmental benefit.

But in my view, on a planet of seven billion people, we must be far more careful when we produce and consume goods and services. We should not reduce consumption of the goods and services we believe to be central to our quality of life, but need to develop different types of "low-eco-impact" economic consumption. The science of Earth observation can help us understand the effects of humans on the planet. More advanced technologies, such as closed-system engineering, can help us increase production while reducing environmental effects.

METRICS

I raised the issue of sustainability metrics when discussing organizational change, but we should examine this issue both inside and outside of organizations. At the (private or public) organizational level, sustainability can be seen as a conceptual framework for assessing the physical elements of an organization's material inputs, work processes, outputs, and outcomes. This assessment requires transparent, reliable, valid, and auditable measures. In the construction of sustainability metrics, it is important to establish a baseline—"the point in the scale below which the thing being measured is unsustainable, and above which it is sustainable" (Dilworth et al. 2011, 23). Measurement is important because, to once again paraphrase management guru Peter Drucker, you can't manage something if you can't measure it. Measuring sustainability and disclosing an organization's impact on the environment helps make the intangible benefits and risks related to these issues more concrete. In the public sector, sustainability indicators and reports give cities an opportunity to assess whether their initiatives are having beneficial effects (Maclaren 1996). Without measures, you cannot tell if your management actions are making the situation better or worse.

These organizational-level measures can then be aggregated by industry or by geography with the ultimate goal of being either included in a nation's GDP or placed in a new indicator of a city, state, or nation's progress toward a renewable, sustainable economy. The state of measurement

in this field is relatively primitive, and we have a long way to go. According to a report by the International Institute for Sustainable Development:

> To date, voluntary sustainability standards have grown principally as a
> result of market forces. Whether motivated by environmental, labour, sup-
> ply, reputational or other risks, sustainability standards have been driven
> by market imperative. Reliance on the market for their existence has
> resulted in the development of initiatives with a high degree of flexibility
> and pragmatism—hallmarks of the voluntary sector more generally. . . .
> While this has helped foster innovation within the sustainability standards
> sector, it is also leading to reduced ability for clear market communication
> and strategic policy intervention. (Potts et al. 2014, 323)

There are some common resources and frameworks that companies
use to measure and report sustainability activities. The Global Reporting
Initiative's Sustainability Reporting Guidelines "offer reporting principles,
standard disclosures and an implementation manual for the preparation
of sustainability reports by organizations, regardless of their size, sector or
location" (GRI 2015). B Lab awards the B Corporation certificate to organizations, ensuring that they meet certain standards of transparency, accountability, sustainability, and performance (CSRwire 2016). The Sustainability
Accounting Standards Board (SASB), a group that develops accounting
standards for sustainability, bases its framework on principles of environmental, social, and governance factors (SASB 2013). The Dow Jones Sustainability Index (DJSI) is a global sustainability benchmark that tracks the stock
performance of the world's leading companies in terms of economic, environmental, and social criteria (RobecoSAM 2015). But despite these services
and other activities related to sustainability metrics, there are still difficulties
in measuring, reporting, and assessing the impact of sustainability.

I am reminded about the time in the United States when every corporation self-reported its financial condition, making a true and fair public
financial market difficult. After the market crash of 1929, during the New
Deal era of the 1930s, the Securities and Exchange Commission began the
process of regulating corporate financial reporting, leading to the development of generally accepted accounting practices. When coupled with
professional and regulated auditing, it is now possible to have some faith

in corporate financial reports. We need sustainability metrics to follow the same course, and we need to develop generally accepted sustainability metrics. Before we can do that, however, we need consensus on what to measure and report, which is no easy task. As a survey of more than 3,100 managers and executives found: "The difficulty of quantifying costs and benefits of sustainability-related strategies and the difficulty of developing comprehensive metrics for assessing sustainability impacts are two of the most frequently cited obstacles for why it's so challenging to make the business case for sustainability" (Haanaes et al. 2011, 79).

The public sector plays a critical role in advancing and supporting sustainability metrics, measurement, and reporting. It can be useful in mandating and monitoring various forms of sustainability reporting and in guiding the development of specific information that private businesses, as well as public and nonprofit organizations, should measure and communicate externally. Over a dozen countries require some type of mandatory sustainability reporting. In the United States, sustainability reporting is not required, although there are many voluntary efforts at the company, industry, and city levels, as well as other efforts to evaluate environmental, social, and governance issues. To advance global progress on sustainability, the United States must move toward mandating environmental disclosure and sustainability reporting.

SUSTAINABLE FINANCE

For sustainability management to become fully integrated into organizational management, it also must be reflected in financial markets and in the factors influencing the deployment of capital. In my view, there are four key elements of green finance:

- The first is to develop private financial instruments and techniques, along with public policies to help attract capital to the sustainable elements of the economy.
- The second is to identify, for investors, the relative risk of investing in sustainable businesses compared to investing in businesses that ignore sustainability.

- The third is to enable private and public organizations to analyze and understand the financial costs and benefits of sustainable practices in assessing their own organization's cost structure.
- The fourth is the growing market in tradable pollution rights, carbon offsets, and other exchanges of environmental assets.

By "sustainable," I mean organizational actions that reduce the one-time use of resources such as energy, water, and other materials and reduce the impact of production and consumption on ecosystems. In a capitalist global economy, attracting capital and analyzing costs and benefits are everyday and routine tasks. If sustainability elements are ignored—or even worse, trivialized—by the financial community, they have little meaning in the real world. But we are finally seeing growing attention being paid to sustainable finance. Over the past 40 years, investor interest in how environmental, social, and governance criteria affect corporate financial performance has surged. According to a report by the Deutsche Asset Management Center for Sustainable Finance (2016, 9), "for some time now investors have become increasingly aware of the materiality of sustainability issues such as climate change, resource scarcity, labor rights, corporate governance and their implications for broader economic and financial stability. This has triggered a new paradigm in the investment landscape to emerge whereby extra-financial factors are moving more into play."

Sustainability finance can only be fully realized when we have agreement on the definition of sustainability and metrics that uniformly measure the achievement of sustainability objectives. Some of the idealism, mission orientation, and even creativity of the field will be sacrificed in order to bring the practice of sustainability management into routine organizational management and finance.

The chief driver behind sustainability is the need to continue economic growth while reducing the impact of that growth on the earth's natural systems. That leads to a greater concern on the part of organizational managers for the physical dimensions of sustainability—their use of water, energy, and materials and the impact of an organization's production, consumption, and products on ecosystems. As sustainability management practices become more standard, they will change the definition of financial risk, management competence, and organizational performance.

This in turn will impact the world of finance. Commercial and investment banks are playing an increasingly significant role in financing the development of large-scale sustainable products and solutions, primarily through asset finance, tax equity, and green bonds. For example, total new investments in clean energy by commercial and investment banks increased 142 percent from 2006 to 2014, reaching $310 billion (Singer 2015). And green bonds—fixed income securities for which proceeds will be used for projects with environmental benefits—have ballooned in the past few years, with issuances exceeding $47 billion in 2015, compared to $10 billion in 2013 (Morgan Stanley 2016).

The world of finance is both data-driven and unforgiving. Capital flows toward the highest rate of returns and/or the most stable assurances of return. But those markets are influenced by government policy, social norms, and consumer behavior. State, local, and some national governments are beginning to push the transition to a sustainable economy. A reluctant finance community is slowly getting the hint, and the field of green finance is beginning to take shape.

SUSTAINABILITY MISMANAGEMENT

While it is easy to see the trend toward sustainability management, there are plenty of examples of sustainability *mismanagement*. Following are three graphic examples of corporations that did not effectively manage environmental risks:

1. The 2015 collapse of a mining dam at the Samarco iron ore mine in Brazil was one of the biggest environmental disasters in that country's history. The massive dam failure released an avalanche of sludge that killed 19 people, destroyed nearby towns, and traveled more than 400 miles to the Atlantic Ocean. While the cause of the dam's damage remains unclear, initial investigations suggested that it could have been avoided. The dam, which belonged to Samarco, was a joint venture between Australia's BHP Billiton and Brazil's Vale. In March 2016, the three companies involved agreed to spend a minimum of $2.6 billion on cleanup over the next 15 years (Kiernan 2016).

2. The British Petroleum (BP) oil spill in the Gulf of Mexico in 2010 is one example of regulatory capture and regulatory failure. For 87 straight days, oil and methane gas spewed from an uncapped wellhead, releasing more than three million barrels of oil into the ocean. Public agencies that were created to oversee and regulate an industry instead served to advance the interests of companies in the industry. The U.S. Department of the Interior's poor performance as a regulator contributed to the disaster. BP will end up spending $61.2 billion on the cleanup and fallout of the spill because of their mismanagement.

3. Volkswagen's deliberate and systematic effort to violate clean air rules is also an example of management incompetence and failure. By 2016, Volkswagen (VW) had already set aside more than $7 billion to help cover the costs of the management incompetence that led to lying about the emissions discharged from its diesel autos. That number will increase over time.

The central argument for sustainability management is that we do not preserve the environment because we love nature (although we might), but because we need it. Our reliance on natural systems will continue as long as we need to breathe air, drink water, and eat food. Poisoned rivers are poor sources of drinking water. Here in America, we continue our decades-long effort to clean up the legacy of our toxic past. While I wish we'd stopped generating this form of pollution, it is important to note that we have not yet ended some of the practices that got us into this mess. According to the National Wildlife Federation (2016):

> The hard rock mining industry is the single largest source of toxic waste and one of the most destructive industries in the country. Today's industrial-strength mining involves the blasting, excavating, and crushing of many thousands of acres of land and the use of huge quantities of toxic chemicals such as cyanide and sulfuric acid. The mines that produce our gold, silver, copper, and uranium are notorious for polluting adjacent streams, lakes, and groundwater with toxic by-products.

Moreover, outside the United States, in the developing world, we see the pattern being repeated. China, India, and countries in Latin America

and Africa provide countless examples of the same sloppy and misman-
aged industrial practices that we practically invented in the United States.
There is no escaping these costs. It is truly a case of "pay me now" or
"pay me later." The short-term, expedient result of ignoring environmental effects may be greater immediate profit for some, but the long-term
impact is higher costs and lower profit, and many of those higher costs
must be borne by all of us. Many of the companies that made the mess will
be long gone before many of the bills come due. Even when the impact
of toxic pollution is immediate, some of the costs to society are hidden in
our rising health-care bills.

Over and over again, we see companies and governments making
short-term decisions to save money, but then see these "pragmatic" decisions costing more money when decisions must be reversed: Fukushima's
inadequate sea wall, VW's deceptive software, BP's inadequate contracting
in the Gulf of Mexico, GE's dumping of PCBs in the Hudson River, Flint's
water supply system. The list is long and getting longer. We live on a more
crowded planet, and to maintain and grow our economy, we must learn
to be more careful in our use of natural resources.

All over the world, from China to India, and here in the United States,
from West Virginia to the city of Flint, Michigan, poor management is
harming the environment, public health, and everyone's pocketbook.
There are no shortcuts, and the sooner the leaders in our governments
and businesses figure that out, the sooner we can proceed with the real
work of growing our economy without destroying our home planet.

I am betting that human ingenuity and the huge profits to be made in
developing new technologies make them inevitable. Moreover, it strikes
me that the fossil fuel business is on the wrong side of history. Yes, extraction technology has advanced rapidly, but the fuel being extracted must
still be paid for. And even though there is plenty of fossil fuel in the
ground, it is still quite finite. The last time I checked, sunshine was still
free and will be around for a very long time. When the underlying fuel is
free, then price is simply a matter of the cost of technology. As technology gets better, the price of renewable energy will go down. Fossil fuels
have the opposite long-term trajectory: as the fuel gets less plentiful and
more difficult to extract and ship, its price will go up. Of course, when
fossil fuels are replaced by renewable sources, their value will crash and

VOLKSWAGEN'S EMISSIONS TESTING SOFTWARE SCANDAL

In an amazing and brazen attempt to avoid compliance with our air pollution rules, Volkswagen (VW) installed a piece of software that allowed its diesel-powered automobiles to circumvent air quality regulation. Reporting in the *New York Times*, Coral Davenport and Jack Ewing (2015) observed that:

> The Environmental Protection Agency accused the German automaker of using software to detect when the car is undergoing its periodic state emissions testing. Only during such tests are the cars' full emissions control systems turned on. During normal driving situations, the controls are turned off, allowing the cars to spew as much as 40 times as much pollution as allowed under the Clean Air Act, the EPA said.

The hero in this story was the International Council on Clean Transportation, a nonprofit organization that provides research and scientific analysis to environmental regulators, and which found a large gap between pollutants emitted in the lab and emissions detected on the road. The organization reported this to the EPA, which led to the investigation, Volkswagen's confession, and the recall of about half a million cars. VW owners may not be happy with the performance of their cars once the pollution control devices are restored, as these vehicles may lose a lot of acceleration potential when pollution controls are engaged. In all likelihood, some owners will not comply with the recall order, and state-managed emission inspections will continue to be fooled.

It is not difficult to understand why auto company management might attempt to circumvent the law this way. The issue for Volkswagen is the depth of this corrupt attitude in its corporate culture. I imagine corporate management somehow felt justified in taking this action. They may have understood that the health impact of air pollution is real and important, but that it would be difficult to assign causality to one source of pollution. It is true that VW's deception was hard to detect, and it is even more difficult to point to the specific harm caused by this deceit.

But what if the same willingness to take shortcuts and deceive extends to safety equipment like seat belts and airbags? How about the car's brakes? If a car manufacturer is dishonest in one area, what's to keep it from being dishonest in other areas?

The issue for Volkswagen was not simply how to comply with EPA's rules, but what does the company do to eliminate the cause of this deception within its organization? The auto industry has a long history of resistance and, at best, grudging compliance with environmental and safety rules. The car companies complained about seat belts and said that requiring them would harm their business. They complained about the catalytic converter. They continue to complain about gas mileage standards. It is ironic that this crime was detected at the same time that Volkswagen was announcing a major increase in its production of electric and hybrid automobiles. It is clear that one part of the company understands the commercial potential of sustainable personal transportation—but clearly another part of VW could care less.

Volkswagen's deception should be seen as sloppy management. It opened the company up to significant fines. At the time of the incident, Coral Davenport and Jack Ewing (2015) reported: "The Justice Department's investigation [of Volkswagen] could ultimately result in fines or penalties for the company. Under the terms of the Clean Air Act, the Justice Department could impose fines of as much as $37,500 for each recalled vehicle, for a possible total penalty of as much as $18 billion."

While the penalty will probably not reach $18 billion, any penalty is bad for the bottom line and for the organization's reputation. Taken a few steps further, a modern sustainability manager would take care to ensure that any use of resources was fully thought through and all environmental effects were minimized. The absence of concern for sustainability is an indicator that Volkswagen's level of management competence needs improvement. Perhaps this incident will stimulate the company to improve. At least the company seemed contrite. Shortly after the software deception was discovered, Martin Winterkorn, Volkswagen's CEO, apologized for the company's misconduct. They also stopped sales of new and used cars containing the deceptive software.

YBP Library Services

COHEN, STEVEN, 1953-

SUSTAINABLE CITY.

Paper　245 P.
NEW YORK: COLUMBIA UNIVERSITY PRESS, 2018

AUTH: COLUMBIA UNIVERSITY.

LCCN　2017012419
　ISBN　0231182058　　Library PO# GENERAL APPROVAL

	List	26.00 USD
5461 UNIV OF TEXAS/SAN ANTONIO	Disc	10.0%
App. Date 02/07/18　　URB.APR　　6108-11	Net	23.40 USD

SUBJ: CITY PLANNING--ENVIRONMENTAL ASPECTS.

CLASS HT166　　DEWEY# 307.3416　　　LEVEL ADV-AC

then their price will go down. More likely, a small amount of these hydrocarbons will retain value and be used to manufacture plastics and other materials that take advantage of their unique properties.

CONCLUSION: THE PROGRESS IN BRINGING SUSTAINABILITY MANAGEMENT INTO ORGANIZATIONAL LIFE

Every day we see examples of corporations, nonprofits, and government agencies taking steps toward sustainably managing their operations. Food companies are recycling their waste, auto companies are preparing for the move to electric vehicles, and real estate developers are building structures that are designed to use less water and energy. Of the top 10 consulting firms in the United States, five of them have "sustainability" as one of the major services they offer to companies (Vault 2016). While the political dialogue stubbornly clings to its insistence that we cannot protect the environment while growing the economy, the economic reality on the ground proves that the opposite is true.

Many businesses are seeing the opportunities in the green economy. New technologies, new services, new knowledge, and new jobs are emerging. In California, for example, the solar industry has grown rapidly, and the state became the first to surpass the 75,000 jobs benchmark (Bebon 2016). The old, clunky, pollution-belching smokestack once represented economic might in the twentieth century. The twenty-first century version is a passive solar-designed building with a park view, with space leased to companies that develop smartphone applications, and ride-sharing services. The high value-added parts of our economy are services, ideas, and software; manufacturing is more automated and produces less wealth. Reducing the costs of energy, water, and other materials is occupying more creative effort than ever before. This drive toward a green economy is not entirely based on idealism but has its roots in a desire to be profitable while protecting the planet.

Why doesn't the reality of this movement receive much attention? In part, because it is good news and people would rather watch reports of a natural disaster than of a natural wonder. The media is a business and its job is not to accurately portray the world we live in, but to make money.

The world we see through the news media is not the world we experience firsthand. The most important changes in how we live largely go unreported, while conflict, murder, and mayhem crowd what we used to call the airwaves. The day-to-day life of families and friends only makes the news when we are forced to respond to tragedy.

The smartphone and global communications are probably the technologies that have had the most impact on human behavior of any inventions in the past 50 years. People spend much more time communicating with each other and sharing facts, events, photos, and perceptions. This is not newsworthy, but like the move toward renewable energy, it is a central part of our contemporary reality. Unfortunately, the media is an unreliable source for understanding the world we live in. News reports must be mined for their bias and for the reason a story managed to achieve agenda status on a crowded media menu. The progress we are making toward sustainability is not always easy to see and is often contradicted by movement away from sustainability. The picture is complicated and sometimes contradictory.

While the transition to a renewable resource–based economy is well under way, there remain plenty of unsustainable practices and businesses in the world. And the people who benefit from those practices and businesses are not shy about defending them. This ranges from the aggressive fossil fuel advocacy of the Koch brothers to the recent lobbying of the plastic bag industry that successfully convinced the New York State Legislature to overturn the New York City Council's recent effort to reduce the use of these bags.

The movement away from fossil fuels is not without victims. Writing about the decline in Wyoming's coal industry and the simultaneous rise of that state's wind business, Coral Davenport (2016) observed: "The new positions and financial opportunities offered by wind and other new-energy industries are not replacing all the jobs going up in coal smoke." Not only are the numbers of jobs smaller, but the skill base is different. While these transitions are inevitable in a capitalist society, government programs and policies are needed to ensure that the victims of this transition receive the help they need to find meaningful employment.

Even among those whose jobs are not under threat, we still see widespread evidence of resistance to sustainability. Opposition to complying with air pollution rules appears to have been embedded within Volkswagen's organizational culture, permitting thousands of vehicles to be sold with software designed to fool inspection tests. This resistance to good practice is far from rare. Many people still accept the idea that we must trade off environmental quality and economic development. But this view is starting to become the exception and not the rule.

In a blog post in 2015, Mindy Lubber, president of the sustainability nonprofit Ceres, noted that American businesses supported the greenhouse gas reduction goals of President Obama's Clean Power Plan in part because they saw the opportunities that plan compliance would generate. Moreover, she observed that the move to reduce greenhouse gases predated government regulation. As Lubber (2015) observed:

> At least 43 percent of Fortune 500 companies have already set targets to reduce carbon pollution, improve energy efficiency and procure more renewable energy. Over half of the Fortune 100 are publicly disclosing their climate and energy-saving targets—and they've collectively reduced carbon emissions by 58 million tons, while saving $1.1 billion annually by doing so. These CO_2 reductions by corporations are the equivalent of retiring 15 coal-fired power plants.

While these moves indicate progress, remember that in 2015, 57 percent of the Fortune 500 companies did not set carbon reduction targets. That may require government regulation before it happens. The transition to a renewable economy will not be instantaneous. It will be a matter of two steps forward and one step back, and for some folks it is simply coming too slowly to save the world. Unfortunately, while the pace now under way may not be fast enough, it will have to do. It takes our organizations a long time to change. Humans are ingenious and creative, but we are also creatures of habit.

But as new technologies are developed that are less destructive to ecosystems, their adoption may begin slowly but may also pick up momentum quickly once they've passed the tipping point of popular acceptance. Cell phones, GPS, and Bluetooth technologies are examples of technologies that were adopted gradually, but increased speed as they became more common. We are seeing a similar phenomenon with rooftop solar cells, solar water heaters, and the Tesla electric car.

While the need to maintain the planet's resources may have begun the move toward sustainability, the new set of constraints posed by ecosystem needs is also generating business opportunity. For example, consulting firm A.T. Kearney (2016) states on its website that "by adopting a sustainability mindset, businesses have an unprecedented opportunity to win customer support, engage employees, and gain competitive advantage while improving lives, communities, and our world." In June 2016, a group of businesses in Minnesota started to work together to push for the development of a circular economy. According to Jessica Lyons Hardcastle (2016) writing in the *Environmental Leader*:

> Dow, 3M and Target are among the 25 major companies and organizations that have launched an initiative to promote the circular economy. The Minnesota Sustainable Growth Coalition says adopting circular economy principles—where raw materials are extracted and made into products that are designed and manufactured for reuse and remanufacturing or recycling—will uncover business growth opportunities and drive innovation. Circular economy principles also promote better waste management by sending less material to landfills.

It is easy to dismiss the moves of all of the Fortune 500 companies to reduce greenhouse gases or these Minnesota companies pursuing the circular economy as a fad or a public relations effort. Clearly, none of these activities are considered major developments by the news media. But they are important signs of progress now under way.

The transition to a sustainable, renewable resource–based economy will take decades to complete. It will require a more sophisticated partnership between government and the private sector than the U.S. federal government seems capable of undertaking. Fortunately, many of our city and local governments seem more adept at forging these relationships. Ideology is less important at this operational level of government, and most decision making focuses on tangible projects and programs rather than symbolic policies and positions.

The change we need will be given operational meaning by the organizations we work for. Just as they incorporated financial reporting, performance measurement, customer relations, employment law, social media

marketing, and many other elements into their standard operating pro-
cedures, so too will they need to incorporate a concern for the physical
dimensions of sustainability: the use of energy, water, and other materials;
recycling and designed reuse of finite materials; and efforts to reduce the
environmental impact of organizational outputs. These changes will be
slow and steady and, like the tortoise, may not attract much attention until
it finally passes the hare.

5

THE ROLE OF POLITICS AND PUBLIC POLICY IN BUILDING SUSTAINABLE CITIES

The development of sustainable cities depends on broad support for the goals of reduced environmental impact and on the partnership between the public and private sectors to build the infrastructure needed in sustainable cities. This chapter begins with a discussion of public-private partnerships and their relationship to the emerging global economy. Despite political events in 2016 such as the British referendum to leave the European Union or Donald Trump's "America First" presidential theme and election, the global economy is here to stay. Technology and competition have created the global economy, and governments will find it very difficult to turn back this tide. Partnerships between the public and private sectors and the emerging global economy are the two key contextual elements that are central to understanding the causes of the drive toward urban sustainability.

I then turn to a discussion of the growing level of environmental awareness in the United States, particularly among young people, and the issue of political salience. I believe that many analysts do not understand how the environment works as a political issue, so I will discuss the politics of the environment. The analysis focuses on the importance of local-level sustainability to building political support. I find that when environmental politics leaves the symbolic and abstract discussion at the national and global levels and turns to local issues such as what do we do with the garbage and how do we deal with traffic, support for the goals of urban sustainability grows. I then ask about the dynamics of sustainability politics and discuss how some local-level sustainability issues become nonpartisan.

I will then discuss public opinion and environmental values and how they influence the politics of urban sustainability before turning to a very conservative strain of support for local-level sustainability, the "not in my backyard" (NIMBY) syndrome. The final section of the chapter speculates on the future of urban sustainability. I tie the development of the sustainable city to trends in lifestyle preferences including wellness, diet, physical fitness, and a general concern for human-environment interaction.

PUBLIC-PRIVATE PARTNERSHIPS AND THE GLOBAL ECONOMY

While private organizations will do most of the work needed to transition to a sustainable, renewable resource–based economy, an active and sophisticated government will be required to provide the infrastructure and regulatory framework needed to bring about the sustainable city. Government must create an even playing field that reinforces the tendency toward sustainability and must partner with the private sector to make these changes feasible and real. Public policy cannot simply be symbolic statements and meaningless pronouncements. It has to stimulate changes in organizational and individual behavior. It also has to provide the public goods—transit, water, waste management, parks, smart grids and so on—to permit urban dwellers to minimize their personal environmental footprint while maintaining a high quality of life.

The fundamental role of government in building sustainable cities is to create the infrastructure and the regulatory environment needed to support human activities that pollute the environment as little as possible. The U.S. Department of Transportation's $40 million Smart City Challenge program, a competition that pledges $40 million to one city to integrate innovative technologies into the transportation network, was made possible by collaboration with private partners, which leveraged $500 million more in funding (U.S. DOT 2016). At the local level, this will also involve sustainability planning, zoning, building codes, and other means of coordinating action and influencing private-sector behavior. For example, New York State's Clean Energy Fund is a 10-year, $5 billion effort to support clean energy market development and innovation, which leverages private funds and encourages private investment in clean energy (NYSERDA 2016a).

The action required to transition off of fossil fuels and other single-use resources requires a sophisticated partnership between the public and private sectors. By sophisticated I mean a partnership between adults seeking both self-interest and mutual shared interest with a minimum of ego, ideology, and group think. There will be some instances when the work that needs to be done—for example, basic research or infrastructure finance—will require federal funds. There will be other instances when the tax code or other incentives will be needed to attract private capital and companies into the market. And there will be even more instances when government action is not needed and the best thing government can do is get out of the way and let the private sector act. By "sophisticated partnership," I mean one that is guided by results-oriented pragmatism rather than symbols, turf, power, and ideology.

Which leads to a promising step that Bill Gates took back in 2015 when he announced that he would develop a multibillion-dollar fund to pay the costs of researching new renewable energy technologies. His fund is a public-private partnership that will be fed by a group that spans more than two dozen public and private entities, including national governments, billionaire philanthropists, investment fund managers, and tech CEOs. According to Gates, "private companies will ultimately develop these energy breakthroughs, but their work will rely on the kind of basic research that only governments can fund" (Wattles 2015). Renewable energy is an issue that requires a technological fix, and a global research fund is exactly what is required. When combined with growing public and corporate understanding of the risks of environmental damage and the advantages of sustainability management, it gives me an additional reason to believe that we can continue to grow the global economy without destroying the planet.

Partnership is needed because it ensures that the work will not be dominated by government or by private organizations but require a cooperative relationship between both sectors. It is also needed because for ideological reasons, U.S. government staffing has been held down to convince people that government isn't growing. In 1966, we had about 2.6 million civilian federal workers; in 2015, the number is about the same. Federal workers were about 4 percent of our workforce in the 1960s and about 2 percent today. In the 1960s, we had about 180 million people residing in this country; today, it is more than 320 million (Norris 2013). Of course, government has grown over the past 50 years—some of that is simply a result of population growth—but all of the growth of the federal government has taken place in private firms under contract to the government. Over the past half-century, we have forced government to manage under tight human resource constraints.

The positive side of these constraints is that government has been forced to become more efficient and better able to manage organizational networks. However, the downside is that talented people drawn to public service avoid the federal government. A *Washington Post* article in 2014 stated that "federal shutdown, furloughs and pay freezes in recent years have eroded the attraction of working for the government" (Rein 2014). That year, 7 percent of the federal workforce was under the age of 30, the lowest share in a decade. We need a capable government to form a partnership with the private sector and lead the transition to a renewable resource–based economy. We need competent public leadership if we are to reinforce and rebuild our decaying infrastructure to enable it to adapt to the needs of urban sustainability. The size of government should certainly be controlled, but a starvation diet serves no one.

The global economy is rationalizing and lowering the cost of production throughout the world. But in the developed world it is having a devastating impact on middle-class manufacturing workers. While installing solar panels, operating energy-efficiency equipment, and building microgrids all create new opportunities, they don't employ the number of people that the industrial-era factories and mines employed. While these may help the few coal miners willing to move to the places that have those opportunities or are able to obtain additional education, they will not help the communities that once relied on the coal industry for jobs and money. Community redevelopment requires that new industries be attracted to replace the old ones that are in decline. Creating the conditions to attract new businesses is more art and craft than science. It requires a public-private partnership that is often blocked by our intense but outmoded political ideologies. As cities become sustainable, their economic foundation will become more brain-based than oriented toward the manufacturing of products. But the issue of worker displacement must be addressed. Workers who lose jobs need to be retrained for new jobs, and care must be taken to ensure that the new jobs are not characterized by downward mobility. None of this is easily accomplished, but addressing this issue requires public-private partnership. It also requires government subsidies. It is not the private sector's job to worry about displaced workers: public policy must address the issue by providing direct employment or incentives for private-sector retraining and employment.

The transition to a global economy is well under way, and the transition to a sustainable economy has begun. The human cost of this transition should not be ignored and requires determined and creative solutions. In my view, that does not mean we need big, bureaucratized government programs, but it does mean we need an assertive state and local government working in partnership with the private sector.

GROWING ENVIRONMENTAL AWARENESS AND POLITICAL SALIENCE

The transition to the sustainable city and to the renewable resource–based economy will require political will and inspired leadership. That in turn will require a base of political support for protecting the environment. Many of us feel a sense of urgency when we think of the steps needed to preserve our planet, but governments attempting to maintain economic well-being and stay in power continue to balance change with stability. I think that more and more people in power understand the need to address climate change, ecosystem protection, and toxics in our production processes. But their political survival and, indeed, our peace and stability depend on a well-functioning economic system. Slow, yet significant change is under way. We see it in the replacement of the water pipes in Flint, Michigan, where 8,000 pipes in its drinking water system are slowly being replaced. One can see it in the U.S. Environmental Protection Agency's designation of the Passaic River as a Superfund site and a planned $1.38 billion, 10-year cleanup announced in 2016 (Remnick and Rojas 2016). We see it in the latest Five-Year Plan for economic development in China. As Brian Spegele reported in the *Wall Street Journal* in 2016:

China will cap annual energy consumption at 5 billion metric tons of coal equivalent by 2020, as Beijing pushes to control the use of resources and curb greenhouse-gas emissions. The cap, detailed . . . in a draft of the government's 13th Five Year Plan economic blueprint, comes as Chinese leaders seek to tackle wasteful resource usage and industrial overcapacity in the world's second-largest economy. It marks the first time a hard energy-consumption cap has been enshrined in a five-year plan. (Spegele 2016)

These are all indicators of a shift in policy direction. In Michigan, everyone knows that Flint's water infrastructure must be restored to safe functioning. In China, everyone who breathes knows that coal consumption must be reduced. In New Jersey, the Passaic River has been a dangerous and disgusting liquid toxic waste dump for more than half a century. The process of change first begins with a change in the direction of policy. This follows a process of research, analysis, communication, and finally understanding of the problem at hand.

Over the past half-century or so, the movement to protect our planet has grown even in the face of important competing needs. Momentary setbacks have always been followed by further progress. China built coal-fired power plants at a ferocious rate but is now moving to transition off of coal. Michigan's government recognized that they made a catastrophic error with Flint's water, and they moved to remedy the situation. Moreover, the visibility of Flint's toxic water crisis increased awareness of the issue of toxics in our water, air, homes, and land. We are slowly addressing problems like climate change and toxics. A half-century ago these problems were barely on the political agenda; today, they are taught in our schools, reported by the media, and acted on by businesses and government.

Environmental issues have always been difficult for politicians and some journalists to understand. In response to poll questions, people often rank other issues higher than environmental issues. For example, while 47 percent of Americans ranked the environment as a top policy issue in a Pew Research Center poll in 2016, strengthening the economy (75 percent) and defending from terrorist attacks (75 percent) were at the top of the public's priority list (M. Anderson 2016). Measuring the public's views of specific public policy issues is more complicated than pollsters want to admit. The salience of a political issue is not a one-dimensional phenomenon. It does not operate like an on-and-off switch. An issue may have tremendous latent appeal but is not a focus of attention because other issues are crowding it out or because in the nature of the issue attention cycle, people get tired of paying attention. According to researchers from the Ross School of Business, the likelihood that environmental damage becomes salient in the public eye depends on both the quantity and visibility of that damage (Heyes, Lyon, and Martin 2016). For example, in a study on what motivates cities to participate in voluntary programs such as ICLEI's Cities for Climate Protection campaign, researchers Zahran and colleagues (2008) found that previous casualties from natural disasters significantly increased the odds that a city would participate. Another critical factor influencing the public's policy priorities is their judgment on whether or not government is making reasonable progress on the issue. For example, the issue of public safety in New York City was a top-10 issue throughout the 1990s; the crime rate was high, and people wanted government to get it under control. The issue still remains very important to the average New Yorker today, but with crime relatively low, it does not rank high on the list of "top 10 political issues." Let the crime statistics spike upward and you can expect that the latent power of this issue will reassert itself. I suspect that environmental issues act in a similar way overall. Today, people notice that the air and water are cleaner than they were decades ago but don't often consider them urgent issues until they are threatened.

There is a general sense of threat to the planet because environmental issues have moved from the periphery to the center of the political and policy agenda in the United States and throughout the world. In the 1970s, the environment was a fringe issue of great importance to a small number of people. Today it is a central, core issue of local, state, and national governance. When the American president meets with heads of state, climate is a key point of discussion. Even a president like Donald Trump is unable to avoid the issue of climate change because its importance is so fully accepted by the entire international community. We have begun the transition to a renewable resource–based sustainable economy. The transition will not always be easy, and whenever we navigate the rough spots we can expect to stimulate political dialogue, controversy, and conflict.

Environmental protection becomes a highly salient political issue when government fails to deliver effective programs to protect air, water, and land, but once those programs are in place the urgency of the issue recedes. However, while the sense of urgency is reduced, it is replaced by a shift in values and in a changed awareness of how the world works. When people experience a damaged environment, it changes their view of the world. They understand ecological interconnectedness—or in Barry Commoner's memorable formulation: "everything must go somewhere." This has nothing to do with environmentalism or ideology. People know that we are stressing the planet's finite resources. Young people know it more than old people, because they have grown up and been educated during the environmental era. Between their college debt, the prospect of downward mobility, and the impact of sea-level rise, the millennial generation is coming of age with a new understanding of how the world works and the nature of that world. It leads to a high level of support for environmental sustainability.

The public likes the idea of shifting to renewables but does not see the transition as a particularly high priority. They do not share the sense of urgency expressed by experts in ecology and climate science. Climate change, biodiversity loss, ecosystems damage, and environmental effects are not always visible and are often difficult to fully understand. We introduce new technologies into our production processes and waste cycles before we understand their impact on the environmental systems we rely on. When effects are obvious and visible they achieve political salience, and we are moved to act. Polluted drinking water in Charleston, West Virginia, radioactivity near Fukushima, Japan, and air quality in Beijing, China, are easy to see, sense, and act on. Other forms of environmental damage must be modeled, projected, or imagined and have lower political impact.

THE IMPORTANCE OF LOCAL-LEVEL SUSTAINABILITY TO BUILDING POLITICAL SUPPORT

For government to facilitate the transition to urban sustainability, we need updated environmental laws and sufficient resources to enforce them. In 1972, we decided we needed national water quality standards, and the Federal Water Pollution Control Act amendments (also known as the Clean Water Act) were passed to ensure clean water, despite then President Nixon's veto. Subsequent national law focused on safe drinking water. If states and cities try to evade their responsibilities to implement water quality rules, the federal government is required to step in and protect the public. The structure of law and responsibility is clear and has been in place for more than four decades. During the Flint, Michigan, water crisis in 2015–2016, we saw the water policy implementation chain was only as strong as its weakest link, and the system needed reform and update.

But most of the actual work of government is done at the local level. Cities are important agents for sustainability because of their population size, environmental impact, and direct service delivery role. Local governments are responsible for schools, police, firefighting, transportation, land use, water, and waste management—not to mention parades and fireworks. The federal and state governments make policy and collect and distribute revenue, but for the most part, the real work of government is local. Researchers Daley, Sharp, and Bae (2013, 146) stated in a study that at lower levels of government, "problems are more likely to be accurately identified, solutions are crafted at the local level by individuals who understand the political and social culture, and feedback and adaptive management can be more immediate." And according to the United Nations Environment Programme, "local authorities construct, operate and maintain economic, social, and environmental infrastructure, oversee planning processes, establish local environmental policies and regulations, and assist in implementing national and subnational environmental policies. As the level of governance closest to the people, they play a vital role in educating, mobilizing, and responding to the public to promote sustainable development" (UNEP 2000).

City-level sustainability initiatives, such as PlaNYC 2030/OneNYC in New York City or Greenworks in Philadelphia, tend to be integrated into local economic development efforts and often enjoy a high level of nonpartisan support. Many local leaders have come to understand that sustainability drives economic growth. According to the New Climate Economy commission, investing in public and low-emission transport, energy efficiency of buildings, and waste management in cities could generate $17 trillion in savings worldwide by 2050 (New Climate Economy 2015). Green initiatives attract business, tourists, and new residents. People can see and experience local-level sustainability initiatives because they have an immediacy not typically seen at other levels of government. In New York City, you can see the bike-sharing stations, the new bike lanes, and the three types of trash and recycling baskets out on the street. Efforts at energy efficiency can be seen in lower utility bills. Federal or state governments fund some sustainability initiatives, but local governments typically implement them.

At the state and local levels, *Governing Magazine* counted water supply and carbon emissions as two of the top 10 "legislative issues to watch in 2015" (*Governing Magazine* 2015). This is an indication that apart from the strategy considerations of American presidential politics, the basic needs of state and local governance indicate that environmental issues are moving to the center of the political process. These state and local priorities could influence presidential primaries and spill into the national election agenda, although clearly we saw little of this in 2016. Despite 2016's relentless national race to the bottom, efforts to avoid addressing environmental issues may become more difficult in our evolving electoral political life.

While we desperately need U.S. federal sustainability policy, in the final analysis the environmental quality that people experience in their home communities will have the highest degree of political salience. A successful strategy to protect our environment will need to focus on local effects. Once again, the late Speaker of the House Tip O'Neill is proven correct: "all politics is local" (and, by extension, all *environmental* politics is local).

Environmental protection is least controversial when it is focused on effects that are too obvious to miss. According to a Gallup poll in 2015: "Americans express greater concern over more proximate threats—including pollution of drinking water, as well as pollution of rivers, lakes and reservoirs, and air pollution—than they do about longer-term threats such as global warming, the loss of rain forests, and plant and animal extinction" (Jones 2015). While climate change is a critical environmental problem, it does not mean that that the only way to address it is head on. Fossil fuel use causes many other problems: air pollution, ecological damage during extraction, political risk to ensure constant supplies from unstable parts of the world, and the economic costs of rising and/or unstable prices. Policies that focus on these visible effects have proven more politically popular than policies addressing long-term and less visible problems such as climate change.

In the United States, it has fallen to states and cities to facilitate the transition to sustainability, and while many have stepped up and acted, others have not. The cities that implement sustainability plans and the states that enforce environmental rules have cleaner air, better parks, and higher quality of life. The most popular sustainability practices in cities include tree conservation, alternative-fuel vehicle adoption, promotion of bicycle use, water conservation, education, and construction of new buildings using Leadership in Energy and Environmental Design (LEED) standards (Wang et al. 2012, 847). In the long run, these assets will attract people and business in the global economy. But a large part of the country clings to the fossil fuel–based economy. They treasure their SUVs and express a fervent desire to turn the clock back to an America that was simpler, more ordered, and somehow bound for glory. I'm not sure that world ever existed, but nostalgia is a powerful political force. Still, people of all political persuasions like to breathe fresh air and drink clean water. Some may never believe the science of climate change, but they know orange water when they see it and they know it is government's job to keep the drinking water clean and safe. One of the attractions of American cities that continue to be based on a suburban sprawl mode of land use is that housing tends to be less expensive, and many people prefer large private spaces. However, even these sprawling cities are beginning to see solar arrays installed on their rooftops along with electric vehicles charging in their garages.

Some of the sustainability policy we see at the state and local levels encourages energy efficiency and renewable energy. These policies reflect a desire to reduce the cost of energy and the recognition that wasting energy makes no sense. These policies are not fully based on political support for sustainability but reflect a complex array of causes including lobbying by firms selling energy-efficiency materials, equipment, and services. Both New York State and California tax energy use and create a fund that subsidizes energy efficiency and renewable energy. New York City and other local governments are working to green their buildings and vehicle fleets and trying to find ways to operate their service delivery systems sustainably. The dedicated tax revenues are programs that could be extended nationally and have a massive impact on reducing energy waste, modernizing the energy grid, and encouraging renewable energy. While such a tax is unlikely, the benefit is obvious.

The politics of energy efficiency and renewable energy at the state and local levels focuses more on day-to-day practical measures and less on grand, macro-policy statements. In New York, the energy utility Con Edison sends contractors to small businesses to conduct energy audits and suggest steps that they can take—with and without government's help—to save money on their energy bills. While there is often controversy in any action taken by government at any level, these programs have gained traction and appear well on their way to being seen as part of the normal landscape of state and local programs. Sometimes states and cities can serve as laboratories for experimenting with innovative policy approaches, which may inform national policy. For example, when California tightened its gas mileage standards, it influenced the federal government to tighten its federal Corporate Average Fuel Economy (CAFE) standards (Goulder and Stavins 2011).

Extreme weather events also have helped build support for more resilient and therefore sustainable infrastructure. According to data compiled from the National Oceanic and Atmospheric Administration and the insurance company AON, in the three years after Hurricane Sandy in October 2012, the United States experienced 97 extreme weather events that have cost more than $100 million each. These extreme events range from hurricanes and tornados to wildfires and droughts to simply severe weather over a period of a few days. Between October 2012 and October 2015, extreme weather events cost the United States $176.8 billion and resulted in 861 lives lost. The most costly of these include Hurricane Sandy in 2012 ($65 billion), droughts and heat waves throughout 2012 ($31 billion) and 2013 ($11 billion), and severe weather in the Midwest in May 2014 ($4 billion) (AON 2012, 2013, 2014, 2015; NOAA 2016).

In the years since Hurricane Sandy, we have had, on average, more than one $100-million-plus weather disaster every other week. During summer 2016, we saw massive forest fires in California at the very same time that parts of Louisiana suffered catastrophic flooding. When an emergency happens twice a month it is no longer an emergency but a regular event. If we invested $50 billion now, could it pay off in averted costs down the road? Probably. Should we be thinking about framing the issue of climate change as one of protecting the public's safety and money from the effects of extreme weather? Definitely. These facts and the understanding of their impact have built political support for state- and local-level sustainability policy.

WHAT ARE THE DYNAMICS OF SUSTAINABILITY POLITICS AND CAN THEY BECOME NONPARTISAN?

Whenever I hear that environmental protection is a partisan issue, I'm reminded of New York City Mayor Fiorello La Guardia's famous statement that there is "no Democratic or Republican way to pick up the garbage." The provision of clean air, safe drinking water, solid waste management, and flood control are all basic public services that people who pay taxes expect to receive. Environmental protection politics in America used to be less ideological. In 1972, a huge bipartisan majority enacted the Federal Water Pollution Control Act Amendments and then as noted previously upheld it over the veto of President Richard Nixon.

In an interesting poll published in 2015, the Pew Research Center reported that the "GOP's favorability rating has taken a negative turn." Republicans were seen as more extreme and were also seen as less interested than Democrats in "people like me." While the parties were seen as relatively even on issues such as taxation, immigration, and the economy, Democrats were considered much better able to deal with environmental issues than Republicans. Pew found that 53 percent of their sample believed that Democrats were better able to deal with environmental issues while only 27 percent believed that Republicans were better at addressing these issues (Pew Research Center 2015). While that gave the Democrats an advantage, the poll did not examine the importance of particular issues to the public or to the candidates seeking office. We can't tell if environmental issues will be the focus of attention in future elections, but the environment continues to act as a partisan political issue.

The environment holds the potential to emerge as a political issue in future elections in part because it has gone from being a nonpartisan

consensus issue to a partisan ideological issue. The battleground is typically for the heart and mind of the independent voter. Both Pew and
Gallup are reporting that around 40 percent of individuals in the voting
population do not identify with either major party and consider themselves independents. The question is the salience or prominence of environmental issues to independents. An indication of the answer appeared
in a *Washington Post/ABC News* poll in 2015. In an excellent analysis of
that poll, Phillip Bump of the *Washington Post* noted the importance of
intensity of view when measuring public opinion on an issue. The bottom
line for intensity is to ask: To what degree would a candidate's view on an
issue be a factor that decided someone's vote? When assessing the issue
of climate change, Bump (2015) observed:

> According to data from the new Washington Post / ABC News poll,
> supporters of government action are actually more likely to be in the
> litmus test realm. When it comes to [the presidential election of] 2016,
> a full 58 percent of registered voters say that they favor a candidate who
> will take action to fight climate change—and 38 percent of all voters think
> that position is very or extremely important. . . . As you'd expect, there is a
> partisan difference in these responses, but that partisan difference reflects
> the overall split on enthusiasm. Democrats fervently want a candidate
> who supports action on addressing the warming climate; Republicans
> more lackadaisically oppose it. Independents, meanwhile, look more like
> Democrats.

The environment did not emerge as an important political issue in
the 2016 election. It is difficult to relate polls on policy issues to election results, and more difficult in the 2016 election because even though
Trump won, Clinton finished well ahead in the popular vote.

Republican support for environmental issues has decreased since 2004
(M. Anderson 2016). According to polls, 74 percent of Republicans with a
college degree say that the climate change issue is exaggerated (Newport
and Dugan 2015). According to an earlier Pew Research Center poll in 2014:

> Nearly two-thirds (64%) of U.S. adults favor stricter limits on power
> plant emissions to address climate change, while 31% oppose such

regulations. . . . These opinions, however, vary greatly by party. Fully 78% of Democrats and those who lean Democratic back stricter emissions limits, compared with only half of Republicans and those who lean Republican. Tea Party Republicans (including independents who lean Republican) are especially resistant to stricter emissions limits for power plants, with 71% opposing stricter guidelines. (M. Anderson 2015)

Views about stricter limits on power plant emissions vary by gender, age, and education. Women, young people, and those with more formal education are most likely to favor emission limits. In other words, the people looking for action on climate change feel the need for government movement more intensely than those that oppose action. Moreover, independent voters—a growing and younger part of the electorate—hold views closer to the Democrats than to Republicans.

Strategically, this means that candidates may find that raising environmental issues during a campaign can help them appeal to independent voters. Because of the relative indifference of their base, once they are nominated Republicans may find it safe to support sustainability policies without seriously alienating their supporters. Democrats can raise environmental issues throughout their campaign, but it is unlikely that their environmental views will differentiate them from their primary opponents who are also likely to be pro-environment.

PUBLIC OPINION, VALUES, AND THE POLITICS OF URBAN SUSTAINABILITY

While it is far from universal, more and more people understand that we need to sustainably manage our planet's resources and ecosystems. This awareness has been growing for about a century but has picked up momentum in the past decade. Evidence of deep support for environmental protection is an old story in the United States, Japan, and Europe, and as urbanization and economic development grow throughout the world, we are seeing it accompanied by increased worldwide environmental awareness as well.

Reporting from Costa Rica in 2014, *New York Times* reporter Justin Gillis (2014) observed:

> Over just a few decades in the mid-20th century, this small country chopped down a majority of its ancient forests. But after a huge conservation push and a wave of forest regrowth, trees now blanket more than half of Costa Rica. . . . Over time, humans have cut down or damaged at least three-quarters of the world's forests. . . . But now, driven by a growing environmental movement in countries that are home to tropical forests, and by mounting pressure from Western consumers who care about sustainable practices, corporate and government leaders are making a fresh push to slow the cutting—and eventually to halt it. In addition, plans are being made by some of those same leaders to encourage forest regrowth.

The climate change issue is perhaps the most difficult sustainability issue, and it need not be central to addressing many urban sustainability issues. Nevertheless, when an individual understands the climate issue, we find it to be an important indicator of a deeper understanding of the dimensions of the global sustainability challenge. The politics of climate change remains contentious, with Democrats more concerned about the issue than Republicans. What is most interesting about the polling data is that young people are far more concerned about climate change than older people. In a 2016 Gallup poll, 61 percent of young adults ages 18 to 34 agreed that the record temperatures in 2015 are due to human-caused climate change, while 52 percent of adults ages 35 to 54 said the same, but only 39 percent of those age 55 and above believed that increased temperatures were due to human factors (Dunlap 2016). Young Republicans, independents, and Democrats all understand the findings of climate science. While we do not have the consensus we see over issues like water pollution, it appears that over the next several decades it is likely that the environmental consensus we saw in the United States in the 1970s could fully reemerge. When public opinion is united behind an environmental issue, it tends to have great force in the American political system. What has changed in the past several years is that the consensus on climate change has broadened; recent polls show that Republicans in the United States have largely come to understand the nature of the problem.

Education, awareness, and even understanding of environmental issues continue to grow. Efforts to deny or delegitimize environmental science are failing. Young people in particular understand what is happening to the planet. These issues are on the agenda and in our mindset and they are not going away. They are on the agenda because they reflect facts: objective environmental conditions that can be observed. Even if sea-level rise is mainly understood through models and projections of the future, air pollution can be seen and smelled. Traffic congestion and deforestation are facts that can be photographed and communicated by travelers and viral videos. And they are. All of this creates a pressure to act and regulate the behavior of those whose actions degrade the environment. In many cases, public pressure will be enough to force an end to ecosystem destruction, but not in all cases. The dilemma that we face is when there are no villains—when the damage is caused by our very way of life. That is the case with the use of fossil fuels. Energy is needed in virtually all elements of the global economy. The only way to end the damage caused by fossil fuels is to find a replacement that is cleaner than, cheaper than, and as convenient as fossil fuels.

NIMBY POLITICS: THE NOT IN MY BACKYARD SYNDROME

One element of the politics of urban sustainability is an impulse to simply sustain what is and avoid all change. People become antidevelopment because they want to protect their way of life and the value of their homes. As our communities have gotten more crowded and transportation more congested, it has become more difficult to site major facilities for waste management, water supply, sewage treatment, mass transit, and pretty much anything from a big-box store to a homeless shelter. We call this the "not in my backyard" (NIMBY) syndrome. It is a predictable and at times appropriate response to inappropriate development, or development that has been undertaken without adequate community engagement and/or impact analysis. The NIMBY syndrome occurs because people do not want to lose what they have, and they do not trust the large and powerful institutions that try to site major facilities near their homes. It is a concept that gained prevalence in the 1960s and 1970s with the growing number of waste landfills, driven by spatial scarcity and health concerns (Hoogmartens, Dubois, and Passel 2016). But the NIMBY syndrome is not a natural phenomenon; it is a social construct that needs to be addressed.

In 1980, when working as part of the team developing the Superfund toxic waste cleanup program, I was assigned the responsibility of developing a community relations policy for government and contractors engaged in emergency and remedial responses at toxic waste sites. Having observed the communications problems at Love Canal where the community and government constantly clashed on the monitoring and cleanup of a dangerous toxic waste site, the U.S. Environmental Protection Agency (EPA) was determined to do a better job of community engagement as the agency built the national program. The government's response to the toxic waste disaster at Love Canal started before Superfund was enacted and continued for years after the federal program began. Eventually, many of those living near the canal were evacuated, but the process of understanding the risks at the site, explaining it to the people living in the neighborhood, and then cleaning up the site was a long and drawn-out drama of many trials and many errors.

Over the past several decades, we have seen a range of innovations that have facilitated other types of emergency response actions and the siting of infrastructure and development. In the early 1980s, the North River sewage treatment plant on the Hudson River near Harlem was bitterly opposed by local elected officials and community activists. The final design of the plant was adjusted to meet community concerns. It included additional air filtration equipment, and the roof of the plant became Riverbank State Park. It is a beautiful facility with playing fields, an ice-skating rink, public spaces, an Olympic-size swimming pool, views of the river, and other amenities. While the plant siting raised serious environmental justice issues, the state park provided the community with much needed recreational facilities.

In recent years, we have also seen the evolution of "community benefit agreements" between community groups, public officials, and developers. At Columbia University (where I work), a 50-page community benefits agreement was developed that committed Columbia to $150 million in benefit payments to community institutions and nonprofits and to heavy use of minority- and women-owned businesses in constructing the university's new campus in Manhattanville, or West Harlem (Fisher, Zients, and Donnelly 2015). While any new development will have supporters and opponents, the goal in the sewage treatment and campus projects was to work with the community to make development projects more acceptable.

THE FUTURE OF URBAN SUSTAINABILITY

Still, environmental issues inevitably find their way onto the political agenda. Governments are expected to protect people from danger, and pollution's impact on health ensures that environmental damage periodically achieves a high priority on the political agenda. In many cases, the issues are framed negatively in an effort to frighten people into action. That can and has worked in the past, but when the solutions called for by environmental activists are wholesale changes in lifestyle, rather than regulation requiring the use of a cleanup technology, I suspect the strategy will tend to backfire. The fact is that people like their mobility, air conditioning, Internet, and all the toys of modern life. People living by the beach do not want to move to the mountains—and even if they do, they would then need to fear mudslides and forest fires. People do not want to give up their way of life, and asking them to do so is a losing political strategy. But in the case of urban sustainability, the change in life is often seen as a positive move to a more convenient and stimulating lifestyle. The move can help protect the planet, but that is not its purpose. Sustainability becomes integrated into the basic fabric of urban life.

What is needed politically and in reality is a positive vision of a sustainable society. In the case of this country, it will need to be built on the traditional values that have always attracted people to America: freedom, rewarding individual achievement, a love of the new and novel, innovation, and acceptance (even if reluctantly) of other people, cultures, and lifestyles. We may end up living in smaller and better-designed personal spaces along with increased access to more interesting and beautiful public spaces. More of us will spend more of our time in cities and towns. Some of our personal transportation may be replaced by mass transit or Uber-like shared transport. Our diets will continue to change; our engagement in physical fitness, health care, wellness, education, and electronic media will increase. And we will pay more attention to the source of our energy, food, and water and will look to ensure that it is renewable and free of toxics. We will pay more attention to where our garbage goes and think about how to make sure that our waste does not go to waste.

Americans are paying more attention to how the outside world influences their individual well-being and the well-being of their children. This is causing them to ask questions about what is in their food, their air, and their water and how their lifestyle choices impact their physical and mental health. This general trend is causing many people to reexamine their lifestyle choices. This is a gradual and almost unseen trend, but it is seen in data such as the declining rate of auto ownership by young people and the reurbanization trend we are seeing in some parts of the United States, which I discussed in chapter 3.

These changes are not simply a temporary fad or a symbolic trend, but a durable element of our changing values. I believe there are two reasons for this shift. The first is the objective degradation of environmental conditions that people can see, smell, or at least view through the media. Whether it is smog in China, drinking water in West Virginia, or the BP oil spill in the Gulf of Mexico, people know these facts. The second reason is related to the growing emphasis on health, nutrition, exercise, and what we sometimes term "wellness." People are paying more attention to their physical and psychological health. In order to succeed in protecting yourself and your loved ones, government must do its part and protect the environment: on a more crowded planet with higher and higher levels of economic consumption, environmental sustainability cannot be assumed, it must be managed.

The next part of this book presents some case studies to provide concrete examples of the progress and obstacles we are seeing as we transition to urban sustainability. The cases will address waste management, transportation, energy, public space, and the sustainable lifestyle. Following each case, I will discuss what the case tells us about the progress toward a sustainable city.

CASES IN URBAN SUSTAINABILITY

P art I of this book presented the basics of sustainability management and discussed the transition to the sustainable city. I hope the issues are clear and the challenges are now defined, although I believe they can be overcome. This part of the book presents a set of examples of urban sustainability to provide a more detailed and granular version of some of the steps required to achieve sustainability. None of these cases detail a smooth and flawless progression toward sustainability. All have imperfect outcomes, but all present progress in reducing environmental impact while maintaining or improving quality of life. The cases include:

1. **Waste Management:** New York City, Hong Kong, and Beijing. This case details waste management practices in these three cities, looking at the political, technological, economic, organizational, and value basis of the waste issue. In growing and densely populated cities, where landfills can no longer be the solution, how will cities dispose of their garbage? The case looks at solutions such as polluter pays options, recycling and composting, waste-to-energy plants and incinerators, and zero-waste policies.
2. **Mass and Personal Transit:** Bus rapid transit in Bogotá, light rail in Jerusalem, high-speed rail in China, and Tesla in the United States. This

case looks at both personal and mass transportation options around the world. Each of the public transit examples here are seen as successful models, however each also faces criticism. The bus rapid transit model in Bogotá is highly congested and expensive for riders. The light rail system in Jerusalem is extremely expensive, and the system is challenged by the Israeli-Palestinian conflict. High-speed rail in China has the potential to improve access to cities, but the high cost is criticized because of China's rising debt. Electric vehicles are growing as a sustainable option for personal transit, and Tesla is changing the game with a lower-cost, higher-range model, but the electric vehicle infrastructure still doesn't support widespread adoption in the United States. Moreover, electric vehicles charged with electricity generated by fossil fuels do not present a real solution to the problem at hand.

3. **Microgrid Development:** New York University, the Power Africa initiative, and Higashimatsushima in Japan. This case looks at how microgrids are being developed and used in three areas. New York University's microgrid proved to be a capable system in the face of Hurricane Sandy and showed that microgrids are powerful tools to protect against natural disasters. The Power Africa initiative, while still largely in development with mixed results, shows how microgrids are being used to increase a region's electricity generation. And in Higashimatsushima in Japan, the government is using microgrids to achieve high renewable penetration in electricity supply to help with energy independence.

4. **Public Space:** High Line Park in New York City, Victor Civita Plaza in São Paulo, Canal Park in Washington, D.C., and Gas Works Park in Seattle. This case looks at four parks around the world, which all were once spaces that were originally used for other purposes. The High Line was a rail line, Canal Park was once a parking lot, Gas Works Park was an industrial manufacturing site, and Victor Civita Plaza was a municipal incinerator. With toxic cleanup and unique designs, along with public-private partnerships, each of these examples shows that a city can transform underutilized space into something that everyone can enjoy.

5. **Sustainable Urban Living and the Sharing Economy:** Uber and Airbnb. This case looks at one aspect of sustainable urban living—the sharing economy. Companies and services such as Uber (ride sharing) and Airbnb (home sharing) are growing in popularity and are seen by some

as a way to create efficiencies in how resources are used. But cities are also learning how and whether they can regulate these types of services. The two examples in this chapter will focus on Uber and Airbnb.

ANALYZING THESE CASES OF URBAN SUSTAINABILITY

I present these cases as an indication that the first steps toward the sustainable city are slowly being put in place. After describing the background of each case, I provide a deeper analysis to understand why they have taken the shape that they have. I look at each case as an issue of values, politics, science and technology, economics and policy design, and management/organizational capacity. This is the same framework I used in both editions of my environmental policy primer, *Understanding Environmental Policy* (Cohen 2006; Cohen 2014).

6

WASTE MANAGEMENT IN NEW YORK CITY, HONG KONG, AND BEIJING

Solid waste management is a challenge for large urban areas around the world. Removing garbage from residential, institutional, and commercial locations in cities is a major logistic and operational task. It is also a critical matter of public health, environmental quality, and economic development. As the world urbanizes, the challenge is becoming more acute. More people means more garbage, especially in fast-growing cities where the bulk of waste is generated, putting pressure on municipal governments to deal with rising costs and environmental effects.

Cities generate roughly 1.3 billion tons of solid waste per year. And waste worldwide is expected to increase to a total of 2.2 billion tons per year by 2025, an increase of 70 percent. The global cost of dealing with all that trash is rising too, from $205 billion a year in 2010 to an expected $375 billion by 2025, with the sharpest cost increases in developing countries (World Bank 2013a). China, in particular, is undergoing an unprecedented increase in waste generation. China's rapid population growth, urbanization, and industrialization have brought with it an enormous increase in the amount of garbage generated, especially in its cities. The quantity of municipal solid waste generated in China's cities is projected to reach 1.6 million tons per day by 2030 (World Bank 2014b).

This case study examines waste management practices in three cities: New York City, Hong Kong, and Beijing. For each of the three cities, I'll briefly describe how waste is managed and look at the solutions to a growing waste problem that each locale encounters. I'll conclude by comparing the practices in these cities and detailing the technical, managerial, and political issues that define the waste management system in each place.

NEW YORK CITY

New York City's millions of residents and millions of businesses, construction projects, and nonresident employees generate 14 million tons of waste and recyclables per year (City of New York 2014). Waste management and removal is handled by two systems: public and private. The public agency—the New York City Department of Sanitation (DSNY)—serves residential buildings, government agencies, and many nonprofit organizations. Private commercial firms must pay private waste carters to remove their solid waste. Spending by both the public and private systems on residential and commercial garbage was about $2.3 billion (the city government's annual budget was $75 billion) in 2014 (Citizens Budget Commission 2014). New York funds public trash collection with general tax revenue—it doesn't charge customers directly for waste collection.

New York City has a long and difficult history in solid waste management. During the twentieth century, the New York City Department of Sanitation relied on a number of landfills and incineration plants for garbage disposal. With plans for new incinerators slowed, first by the Great Depression and then by World War II, the city found itself struggling to meet its waste disposal needs. In 1947, the Fresh Kills Landfill opened on Staten Island, one of the city's five boroughs. However, as environmental awareness grew, public pressure began to mount against incineration and landfilling. Old landfills and incinerators were gradually shut down, with the last municipal incinerator closed in 1992. By the late 1990s, Fresh Kills was the only remaining waste disposal option managed by the DSNY (Earth Institute 2001). Then, in December 2001, the Fresh Kills Landfill was shut down. As efforts to build local waste-to-energy incinerators were blocked, waste export became the only option for New York City. The city council adopted a 20-year plan for exporting government-managed waste, relying on a truck-based system and a combination of local, land-based transfer stations that took the city's garbage and disposed it in landfills, recycling facilities, and waste-to-energy plants in neighboring states and in places as far as 750 miles from the city (DSNY 2006).

Of the 3.8 million tons of solid waste that the New York City Department of Sanitation now collects annually, 14 percent is recycled, 76 percent is sent to landfills, and 10 percent is converted to energy at a waste-to-energy facility (Citizens Budget Commission 2014). Waste transfer in New York City is land-based, expensive, and environmentally damaging, though well organized and operated successfully by the Department of Sanitation. A plan to barge garbage out of the city is slowly being implemented. Relying on waste export systems leaves the city vulnerable over the long run, as both restrictions on waste disposal and its costs are likely to escalate. The environmental costs are also rising. New York City's landfilled trash produces more than one million tons of greenhouse gas emissions, the equivalent of burning 13,500 tanker trucks of gasoline.

New York City's Waste Management Strategy

New York City is currently pursuing several different strategies to improve waste management, including increasing recycling capture rates; encouraging residents and businesses to divert organic material from landfills; and overcoming permitting obstacles related to waste-to-energy facilities. In 2006, the city council approved a Comprehensive Solid Waste Management Plan (SWMP), which aimed to establish a cost-effective, reliable, and environmentally sound system for managing the city's waste. The cornerstone of the city's recycling efforts is its curbside program, which collects paper, metal, glass, and plastic. The plan aimed to reduce the city's dependence on a truck-based export system, to export in a manner that is cost-effective, environmentally responsible, and sensitive to the local communities. It also aimed to export less waste. In 2011, solid waste management became incorporated into sustainability planning under PlaNYC, New York City's comprehensive sustainability plan. One of the goals under PlaNYC was to divert 75 percent of solid waste from landfills by 2030. It aims to reduce annual greenhouse gas emissions by 34,000 tons while diverting 2,000 tons of waste per day from land-based solid waste transfer stations in Brooklyn and Queens to marine transfer stations (City of New York 2014). The city opened a Materials and Recovery Facility at the South Brooklyn Marine Terminal that sorts metal, glass, and plastic. With the opening of this facility, the New York City Department of Sanitation expanded the curbside recycling program to include all rigid plastics—the first expansion of the program in 20 years.

The Department of Sanitation also launched a voluntary residential organics recycling program in parts of Staten Island, Brooklyn, and the Bronx and expanded the public school food waste composting pilot program. The city added more public recycling bins, which by 2016 reached 2,190 bins. Organic waste accounts for almost one-third of the waste generated by businesses, and the majority ends up in landfills. To combat this, the city passed the Commercial Organics Law in July 2016. Under this rule, New York City businesses that generate large quantities of organic waste are required by law to separate and properly process their organic waste (DSNY 2016).

In April 2015, Mayor Bill de Blasio announced the rebranding of PlaNYC to OneNYC, a plan for a strong and just city that includes strategies for growth, sustainability, resiliency, and equity. Under this plan, the city's goal is zero waste by 2030, or at least no waste sent to landfills. Since the release of OneNYC, the city has made incremental steps toward its zero-waste goal. In 2015, the Department of Sanitation expanded curbside organics collection (mainly food waste) by more than 50,000 households and is now serving more than 700,000 residents (City of New York 2016a). The city also launched the New York City Housing Authority (NYCHA) recycling program in 850 public housing buildings. In February 2016, New York City adopted revised commercial recycling rules that make it easier for businesses to recycle. One area that has had mixed success, however, is reducing the use of plastic bags in the city. After a contentious vote, the city council voted to require certain retailers to collect a 5-cent fee on plastic bags, but the program was delayed by New York's state legislature (Goodman 2016).

The city plans to unveil a new Solid Waste Management Plan in 2018 that will shift nearly 55 million yearly miles of the Department of Sanitation's waste-hauling vehicle traffic from trucks to barges and rail transport.

HONG KONG

Like many other developed regions, Hong Kong has seen its waste levels grow as its economy has grown. Since the mid-1980s, Hong Kong's municipal solid waste load has increased by nearly 80 percent, mirroring the city's rapid economic expansion and population growth over that same period (Environment Bureau 2013, 4). Hong Kong is one of the most densely populated urban areas in the world, with an estimated 7.2 million residents, and the city generates more than 18,000 tons of municipal solid waste per day, around six million tons annually (GovHK 2016). The Environmental Protection Department (EPD) oversees all of Hong Kong's waste and is responsible for facilities management, waste reduction programming, and policy implementation. Waste collection and removal is managed by the Food and Environmental Hygiene Department and by private contractors (Yau 2010). The current operational costs for waste collection, transfer, treatment, and land-filling amount to HK$1.4 billion per year (US$181 million) (Environment Bureau 2013, 23).

Domestic waste accounts for 45 percent of total waste disposed at Hong Kong's landfills, which consist largely of food waste. All domestic waste is collected by the Environmental Protection Department and transferred to landfills through a network of refuse transfer stations. Commercial and industrial waste are collected by private waste collectors, with the exception of some industrial companies who deliver their waste directly to landfills for disposal (EPD 2013). Another major source of waste is construction waste, which is generated from Hong Kong's frequent construction and demolition activities.

During most of the twentieth century, Hong Kong managed its solid waste with a combination of landfills and waste incineration plants. In the mid-1980s, health and environmental concerns led to the dismantling of solid waste incineration (Yau 2010). In 1989, the Environmental Protection Department published its first waste management policy, the Waste Disposal Plan (WDP), which called for the development of an extensive network of waste transfer stations and three new, large, rurally located landfill sites to serve expanding disposal needs. At the time, the city operated 13 landfills, which were being phased out or closed. The new landfills were established in three corners of Hong Kong—North East New Territories (NENT), South East New Territories (SENT), and West New Territories (WENT)—and began operation in the mid-1990s. It cost nearly HK$6 billion (roughly US$775 million) to build these landfills, and the operating cost of all three is around HK$700 million (roughly US$90.3 million) per year (EPD 2016a).

At present, landfills are the primary method of waste disposal in Hong Kong. However, growing population and commercial activity has created new pressures on the city's waste management system. The city's existing three landfills, which accept more than 14,000 tons of waste per day (EPD 2014, 3), are expected to reach their design capacity by 2020, possibly earlier if waste disposal levels continue to increase at the current rate (Environment Bureau 2013). While plans to expand the landfills are being implemented, Hong Kong's high-density population and lack of space limit the scope of landfill expansion to increase capacity. Furthermore, the landfill expansions have aroused public controversy among residents, who worry that the expansions would harm the environment and residents' health.

Hong Kong's Waste Management Strategy

Hong Kong has largely focused on recycling as a way to reduce the amount of waste going to landfills. When recycling became a policy focus in 1997, about 70 percent of municipal solid waste was disposed of at landfills, and only about 30 percent was recovered for recycling (EPD 2006). In 1998, the Environmental Protection Department released the Waste Reduction Framework Plan (WRFP), which set out a 10-year recycling implementation program that shifted the emphasis from collection and disposal of waste at landfills to waste prevention and reuse of waste materials. To boost the recycling rate, waste separation bins were provided on the ground floor or in designated common areas in public housing (Yau 2010).

As garbage levels continued to rise, reaching just under eight million tons in 2002, Hong Kong realized it needed to implement more waste reduction measures to keep pace with the growing rate of garbage. In early 2001, the government announced it was reserving land for the development of a recycling park (later renamed "EcoPark") in Tuen Mun. The area would be subsidized to encourage the local recycling industry. The EcoPark began operation in 2007 and now leases 13 lots for the recycling of waste products. In 2005, the Environmental Protection Department published a new 10-year waste management strategy that set out new recycling initiatives, employing policy tools such as the polluter-pays principle (PPP) and various producer responsibility schemes (PRS)

(Ross 2008). The plan also included the launch of the Program on Source Separation of Domestic Waste, which encourages private residential property management companies to provide waste separation facilities on each floor of a building. As of April 2016, 2,055 buildings had joined the program (EPD 2016c).

In 2013, the Environmental Bureau released the "Blueprint for Sustainable Use of Resources," announcing Hong Kong's target to reduce the per capita disposal rate of municipal solid waste by 40 percent by 2022 (Environment Bureau 2013). To meet this goal, the government established additional initiatives aimed at increasing recycling activities, including the Community Recycling Network and the $1 Billion Recycling Fund, which was launched in October 2015 and provides funds to upgrade the operational capabilities and efficiency of the recycling industry (EPD 2016b). Reducing the amount of food waste in landfills is also a central focus in the 2013 blueprint, because food waste both depletes landfill space and damages the environment. About 40 percent of the food in the city goes uneaten, creating about 3,500 tons of unwanted food each day (Environment Bureau 2013, 16). Currently, there is no ordinance controlling food waste management in Hong Kong, but a few voluntary programs are in place. The city also plans to build organic waste treatment facilities for the recycling of food waste (Environment Bureau 2013, 22).

There has been some progress. A private firm, HK Recycles, offers full-service recycling pickup for a monthly fee. And recycling is growing. Between 2001 and 2014, Hong Kong's domestic recycling rate grew from 10 to 23 percent, and today the city's overall municipal solid waste recovery rate is at 37 percent (EPD 2014, 15; Yau 2010, 2443). Still, the impending closure of the city's three key landfills is looming. According to the Environmental Protection Department: "Even if we increase our recycling rate, there is no way we can deal with the thousands of tons of waste generated by our households, restaurants and construction sites" (Allman 2013).

BEIJING

Beijing, the capital of China, is a huge metropolitan area of more than 21.15 million people (Global Times 2014b). Like Hong Kong and many other large and rapidly growing cities in China, it continues to face waste management issues as waste generation exceeds disposal capacity. Currently, the Beijing Municipal Administration Commission (BMAC) in cooperation with the Beijing Municipal Environmental Protection Bureau (BMEPB) oversees waste management while the Beijing Municipal Environmental Bureau of Commerce (BMBC) oversees material recovery. In 2006, the Beijing Environment Sanitation Engineering Group Company was contracted to handle the collection and transportation of municipal solid waste: this company collects waste daily from the streets and transports it to transfer stations and disposal areas (Wang and Wang 2013). In 2006, Beijing spent about US$242.8 million on waste management and collected and transported 149 million tons of waste, 92.4 percent of which was sent to landfills (Zhang, Tan, and Gersberg 2010; Zhen-shan et al. 2009). Transportation is the largest single cost in Beijing's waste management system, because waste treatment sites are located far from the city.

Prior to 1979, Beijing had no waste regulations. Waste was sent to rural areas and dumped without any restrictions, often to provide fertilizer. However, as the waste residue became increasingly non-biodegradable, Beijing searched for disposal alternatives. The quick solution was to build landfills. Beijing's first landfill was built in 1994 and by the early 2000s, 90 percent of Beijing's garbage was sent to landfills (Global Times 2014a). Landfills were considered a good option because of their cost-effectiveness and ability to accommodate large fluctuations in the type of waste. The exact number of landfills servicing Beijing today is not clear, although sources say there are 15 landfill sites and 6 full transfer stations in the city, along with two incinerators and two composting plants, for a total capacity of 15,280 tons per day in 2010 (Wang and Wang 2013). This doesn't include any illegal landfill sites—of which there could be hundreds. According to the Beijing Commission of City Administration and Environment, the city's waste facilities can process about 21,000 tons every day (Jinran 2015). In any case, data indicate that waste generation rates exceed the capacity of disposal plants. The government began to see that the growing level of waste would not be supported by landfills and since 2008 began to place a greater focus on waste reduction and resource conservation, incineration, and, to some extent, composting as a substitute for sending waste to landfills.

Beijing's Waste Management Strategy

Incineration is increasingly seen as an important solution to Beijing's solid waste overflow problem, especially in waste-to-energy plants. There are two fully functioning incineration plants in Beijing—the Lujiashan incineration facility in the Mentougou district and the Gaoantun incineration facility in the Chaoyang district (Wang and Wang 2013). The Beijing government is planning to build more incineration plants: one report noted that there were between 4 and 10 incinerators under construction (Global Times 2014a). However, public opposition to incineration plants is high because of environmental and health concerns. According to media reports, at least six new incineration plants have been postponed because of public opposition (Balkan 2012). Emission controls for waste-to-energy plants tend to be more lax than those for coal-fired power plants, and the fly ash that comes from outmoded incineration processes can be highly toxic. Legally, incinerators can emit nitrous oxide and sulfur dioxide at four to five times the level allowed for power plants (Balkan 2012). Recently, China has begun searching for cleaner alternatives. In February 2016, an investment group controlled by the Beijing municipal government bought Germany's leading waste-to-energy firm, EEW, to gain access to their latest technology. EEW has developed advanced technology for filtering of emissions and is regarded as an industry leader in this field.

In addition, since 1996, Beijing has attempted to practice separation at the source and promised in the city's bid for the Olympics to achieve a separation rate of 50 percent before 2008. According to BMAC, this was achieved by 2007, with 4.7 million people taking part in waste source separation. In 2006, Beijing recycled 1.638 million tons of material. Much of the recycling sector in Beijing is based on an informal system of "waste pickers." By some estimates, there are 300,000 individuals in the city who manually pick up and recycle material, although they are concentrated in only a few areas of the city. These workers go door-to-door to collect plastic bottles or arrive for collection when called. Many low-income citizens make their living collecting plastic bottles. These informal collectors may be responsible for removing as much as 30 percent of the city's waste.

The government has attempted to increase efficiency in separation through targeted pilot programs in some areas. A trial in 2013 that

TABLE 6.1 Summary of Waste Management Practices in New York City, Hong Kong, and Beijing

PARAMETER	NEW YORK CITY	HONG KONG	BEIJING
Population	8.4 million	7.24 million	21.7 million
Area	790 km²	1,104 km²	16,800 km²
Waste generation rate	More than 14 million tons of waste and recyclables per year (OneNYC)	6 million tons per year of solid waste (EPD)	4.13 million tons (in 2006, when the population was 13.33 million)
Spending on waste management ($US)	$2.3 billion annually (2014)	$1.4 billion annually (2013)	$242.8 million (in 2006)
Agency responsible for waste management	New York City Department of Sanitation (DSNY)	Environmental Protection Department (EPD); Food and Environmental Hygiene Department	Beijing Municipal Administration Commission (BMAC); Beijing Municipal Environmental Protection Bureau (BMEPB); Beijing Municipal Environmental Bureau of Commerce (BMBC)
Private partners	The Business Integrity Commission	Not specified	The Beijing Environment Sanitation Engineering Group Company
Year first landfill was built	1947	1960	1994
Major goals	Zero waste by 2030 (OneNYC)	Reduce the per capita disposal rate of municipal solid waste by 40 percent by 2022 (2013 Blueprint for Sustainable Use of Resources)	Incinerate 30 percent of municipal solid waste by 2030 (China's Twelfth Five-Year Plan)

included five million Beijing residents equipped neighborhoods with sets of household trash bins. More than 20,000 volunteers were sent to the neighborhoods to teach and promote recycling practices; however, residents often did not follow the rules. Other recycling innovations have been introduced as well. For example, in 2012, the city introduced a "reverse vending machine" at a select number of subway stations where individuals can insert plastic bottles into a machine and get subway credits in return (Watts 2012). However, the effectiveness of these types of programs—whether focused on outreach and education or on new programs—remains uncertain, especially as many are in their beginning stages and only in select areas in the city. Furthermore, the complicated and often informal recycling system makes it difficult to know how much is actually being recycled and by whom.

CONCLUSION: WASTE MANAGEMENT

It would be difficult to find a more fundamental urban environmental issue than solid waste disposal. It is clear that effective waste management is essential to the development of sustainable cities. Many cities around the world are implementing innovative measures to deal with waste and are increasingly incorporating waste management into sustainability plans. Some cities are setting positive examples through aggressive recycling and zero-waste programs. Cities are reducing food waste with better storage and transportation. They are implementing construction strategies that increase reuse of materials and investing in waste-to-energy technology. Some local policies such as waste disposal fees and other charges are being used to encourage waste reduction. Some cities have banned the use of plastic shopping bags, and some are requiring that stores charge for the use of bags.

The current crisis of landfill capacity in New York City is now more than 15 years old—but the issue is new in large cities in China and those in other rapidly developing countries. The need to remove waste from households is a simple matter of public health, and all cities are learning how to reduce, collect, and manage their waste more efficiently and effectively. The political, managerial, and technical challenges associated with waste disposal in cities are numerous and complex. A collaborative approach between the private sector, local authorities, planners, and developers will need to be taken to ensure cities and buildings of the future are sustainable and will meet the needs of generations to come.

ANALYSIS OF WASTE MANAGEMENT

As these cities transition toward sustainability, it is useful to compare and contrast the causes and effects of waste management practices in New York City, Hong Kong, and Beijing. In my book *Understanding Environmental Policy*, I assess environmental issues from a distinct set of perspectives, or, to borrow from Graham Allison (*The Essence of Decision*), "conceptual lenses." The framework is used here to assess the problem of waste management as an issue of values, an issue of politics, an issue of technology, an issue of economic and policy analysis, and an issue of management and organizational capacity. The waste strategies employed by these three cities provide lessons for other cities that are trying to solve the problem of managing growing waste in increasingly congested and overpopulated cities.

Waste as an Issue of Values

The generation of waste begins with the individual values that shape the consumption patterns responsible for creating tons of residential and commercial garbage each day. The types and quantities of waste that individuals generate are influenced by economic development, lifestyle, and patterns of land-use development. The use of large amounts of packaging material in distributing goods reflects a community's collective values. Exportation of waste is based on a desire to avoid the potential environmental insult of seeing and smelling garbage and on the values that underlie the "not in my backyard" (NIMBY) syndrome. In New York, this value system has kept waste reduction off the political agenda, but this is not unique to New York City; these consumption patterns prevail in all modern, developed economies. Citizens of Hong Kong and Beijing also value the benefits of a "throwaway" society.

There is also a subtle value choice reflected by the public and governing elite's avoidance of the waste issue. Perhaps part of the problem arises from the fact that garbage is physically unpleasant and reminds some of us of our relative wealth in the face of poverty. We discard food and clothing from which the world's poor could derive benefit. Garbage is also ugly and smells bad. We prefer not to think about garbage or where it will end up. Coupled with this attitude is the historic tendency to keep garbage processing as far away from the middle and upper classes as possible. This coupling of convenience-driven consumption with "waste avoidance" are the values underpinning the solid waste management crisis.

Waste as a Political Issue

The value issues described above have created a political climate that makes it difficult for local decision makers to address solid waste issues. At the core of the solid waste issue are local politics around the siting of waste disposal and treatment facilities. Garbage is inherently undesirable, and it is difficult to identify the benefit of serving as the host site for a community's waste. The political antipathy to waste in New York City was evidenced for more than two decades by the local politics of waste in Staten Island; the highest priority for most of Staten Island's elected officials during the 1990s was closing Staten Island's Fresh Kills Landfill. By the time Fresh Kills closed, all of New York City's residential waste was dumped there. With few exceptions, local politicians have accommodated the long-standing aversion toward locating waste facilities in New York City. The politics of waste, particularly the community politics of siting, has been the principal constraint on policy options for managing the waste in New York City.

In Hong Kong, the government's proposal to expand the existing three landfills as a solution to its growing waste problem has been met with intense opposition from communities and politicians alike. There has also been significant pushback from the community with respect to siting of the integrated waste management facility (IWMF) on an artificial site near Shek Kwu Chau, a small island south of Lantau (Ross 2008, 4). In Beijing, the government has placed increased emphasis on incineration as a solution to the growing waste problem, but officials are starting to face similar siting issues despite a very different political environment. Local opposition to the building of incineration plants has grown, and this public opposition has prevented new incineration plants from being built.

Waste as an Issue of Science and Technology

The high population density of any city cannot be possible without a number of technological innovations, including, of course, solid waste removal. The technology of waste incineration has advanced dramatically since the 1960s. In New York City, waste-to-energy plants and other advanced waste treatment technologies, supplied by marine waste transfer stations or by rail transfer, are in all likelihood the most environmentally sound methods of disposing the waste generated by millions of residents and visitors. Similarly, in Hong Kong and Beijing, the construction of a waste-to-energy facility is seen as a necessary component to manage the city's growing waste loads. The incinerator being built at the integrated waste management facility (IWMF) in Hong Kong will use thermal treatment technology, which will reduce the waste volume into compounded ash. Beijing's recent acquisition of Germany's leading waste-to-energy firm demonstrates China's desire to acquire more advanced, clean technologies to address their growing waste problem.

Despite the existence of appropriate and effective waste disposal technology, the politics of siting still dominates the issue as experts are not trusted, and the government sometimes lacks credibility with the public. Science has a "solution" to this problem, but politics makes use of new waste management technology difficult. For example, an alternative to thermal treatment technology is plasma gasification. While current waste-to-energy technology converts trash into compounded ash, plasma gasification converts organic matter into synthetic energy and removes toxic waste from the trash. However, it is an expensive technology to use and a hard sell for politicians to their constituents. If experts credible to the public and interest groups could confirm the improved technology, scientific fact *might* influence the political dialogue. The technology of waste treatment is developing quickly, and it is possible that a proven, clean waste treatment technology might someday be developed that could overcome political opposition to facility siting.

Waste as an Issue of Public Policy Design

Until recently in New York City, waste disposal did not pose a major fiscal dilemma because of low pricing. Now, as costs are rising rapidly, cost-benefit calculations are starting to influence the policy-making process. The city's current financing structure for waste disposal presents some policy challenges. New York uses general tax revenue to cover residential and public waste collection, so residents are unaware of the real cost of waste. However, if disposal costs continue to rise, waste disposal as a public policy issue may be redefined, thus providing legitimacy to the search for alternatives to waste export.

Another aspect of the solid waste dilemma as a policy issue is its regulatory dimension. Local, state, and federal governments in the United States regulate waste disposal, and the immediacy of enforcement makes the regulatory dimensions of this issue relatively straightforward. Hong Kong's government, in contrast, has tended to favor voluntary policies over mandatory ones. Recycling is a key element of Hong Kong's public policy with respect to waste, but the oversight and enforcement of these initiatives, like the waste separation program, generally falls on building owners and residents. Voluntary policies may be less effective than mandates in shifting the public's attitudes toward recycling and other waste reduction strategies. Similarly, Beijing has not forcefully or effectively regulated solid waste management. Effective public policy design is also made difficult by inaccurate or unavailable data on waste. The definition of "municipal waste" is not consistent between cities in China, and there is an overall lack of data and research on landfills and incinerators and associated environmental effects. A major issue with waste-to-energy in China is that these facilities operate under regulations that are much more relaxed than those for coal-fired power plants, creating environmental and health issues. Chinese municipal governments have been testing different types of policies to incentivize incinerator construction, including tax refunds, prioritized bank loans, subsidies, and feed-in tariffs for electricity sale onto the grid (World Bank 2014b).

One partial solution to the problem of waste disposal is a policy that encourages waste reduction. One option for New York City, Hong Kong, or Beijing might be the use of a fee-for-service system for waste disposal. In such a system, charges are levied for all waste pickups, and rates for recycling are significantly lower than those for mixed waste. However, while this type of policy design has proven effective in places dominated by single-family homes, many residents in these large cities live in apartment buildings, making it difficult to connect fees to individual behavior (Cornell Waste Management Institute 2001). A fee might also encourage illegal dumping. Furthermore in Beijing, the informal system of waste pickers might make this type of policy difficult to implement.

Waste as a Management Issue

Removing garbage from residential, institutional, and commercial locations in New York City, Hong Kong, and Beijing is a major logistic and operational task. Private firms remove the waste from New York City's commercial establishments, but waste from the city's residences, governments, and nonprofit organizations is removed by the Department of Sanitation, which employs thousands of people to do the work. In Hong Kong, waste removal is a complex system that involves government workers, contractors, cleansing workers, and those employed by private waste collectors. There are about 400 active private waste collectors, with some also involved in waste recycling and reprocessing operations. In Beijing, there are similarly thousands of people who engage in sanitation work—many who are not employed or managed by the local government.

Recycling and waste transfer/final disposal are two tasks that have presented challenges to sanitation managers in New York City. Because of previous investments in specific types of collection trucks, the Department of Sanitation's collection vehicles must conduct separate trips for recycled paper, glass, plastic, and mixed garbage. One of the reasons that recycling costs more than traditional waste disposal in New York City is that collection trucks often complete their routes more empty than full. Because a route costs almost the same to run with full or half-full loads, the collection cost per ton of recycled waste is quite high. Another major unsolved management dilemma in New York City is the price of long-term disposal and the uncertainty about the availability of waste disposal facilities. Today, the city has contracts with out-of-state landfills and incinerators to accept the city's waste, but the price of disposal continues to rise, and the supply of disposal sites is not guaranteed.

Waste management in Beijing often involves complicated arrangements, with governing bodies that have overlapping responsibilities, and sometimes leaving gaps in responsibilities. Waste collection services vary largely between Chinese cities, and even within different parts of the city. For example, in Beijing, waste collection services in modern, high-rise apartments function well, while collection systems in poorer suburban areas are much more rudimentary (Zhang, Tan, and Gersberg 2010). The overall system has both formal and informal elements. The formal system includes employees paid by local government or businesses who collect and transport waste, while the informal system is much larger in number and includes waste pickers who make money by the sale of the materials they collect. Waste pickers often collect in unauthorized areas—such as at collection sites, where they disrupt operations, or at landfills, where they face health and safety hazards. Landfills are often not well operated, with more than 1,000 unregulated landfills across the city that release toxic pollutants into nearby water and soil.

Waste as a Multidimensional Issue

Our consumption and our associated patterns of waste disposal are based on our system of values. This includes a fast pace of activity and the use of huge amounts of surplus food and other materials. It is possible to imagine a slower and more deliberate journey through our day and a far more careful use of resources. Compare the way your grandmother might have used a chicken to the way we do. Grandma used every part of the chicken for something. In our wealthy modern life, a lot of the chicken ends up tossed in the garbage.

The siting of waste management facilities is about as political as anything can get. No one wants a waste facility near their home. Technology can play a huge role in recycling, burning waste to make energy, or in mining food waste for nutrients and fertilizer. And of course, organizational capacity must be developed to operate the waste management system. While no single factor dominates in any particular time or place, we might expect one or two issues to dominate. One of the reasons that waste-to-energy plants stimulate political opposition is that some plants have outmoded technologies that pollute. If the technology can be demonstrated to have improved, the politics might become less difficult for people trying to site facilities.

7

MASS AND PERSONAL TRANSIT

The sustainable city must have a reliable, accessible transportation system that includes both public and personal transit options. As cities become more dense and populated, they must still provide a system for people to move, live, work, and play. According to the International Association of Public Transport, "Given that land is one of the scarcest resources in cities, aiming for a higher use of public transport could bring the added benefits of allowing urbanised space to serve other functions, which can in turn improve the liveability, or productivity of the metropolitan area" (UITP Observatory of Automated Metros 2015b). In the United States, public transit ridership has grown by more than 20 percent in the past decade (FTA 2016). Each year, Americans make more than 10 billion trips using public transit, including rail, bus, trolley, and streetcar (*Governing Magazine* 2016). Worldwide, metro systems carry more than 160 million passengers per day, a 7.9 percent increase compared with 2012 (UITP Observatory of Automated Metros 2015a).

But public transit will only go so far in solving transportation issues, as a sustainable transit system needs to address personal transit, especially for sprawling urban areas. The number of cars on the road is expected to reach two billion by 2030, with particular growth in China and India, where cars will likely use gasoline or diesel fuel (Jolly 2015). A sustainable transit system will have to include personal transit options that are less polluting than current internal combustion–based vehicles. Electric vehicles (EVs) are one of the solutions gaining popularity currently. There were roughly 1.3 million EVs used worldwide in 2016, compared to just 100,000 in 2012 (Statista 2016).

This case study will examine four examples of mass and personal transit around the world: bus rapid transit in Bogotá, light rail in Jerusalem, high-speed rail in China, and electric vehicles in the United States, focusing on the Tesla.

BUS RAPID TRANSIT IN BOGOTÁ

Bogotá, a city of eight million, is the political and economic capital of Colombia. While a thriving global city, it is also well known for its congested and hard to navigate roads. Though high traffic continues to plague the city, recent investments by both the state and city governments have helped address traffic problems. These include the creation of one of the most extensive bike path networks in the world, the 200-mile-long "CicloRuta." Another recent innovation is "pico y placa" enacted in 1998 by Mayor Enrique Peñalosa. This policy regulates rush hour traffic by placing restrictions on vehicles on the basis of the last digits of their license plate numbers. Perhaps the most well-known traffic innovation in Bogotá is the TransMilenio, the city's bus rapid transit (BRT) system. The TransMilenio, like other bus rapid transit systems, streamlines transportation via dedicated bus lanes, train station–like bus stops, and feeder buses that connect this central network to the peripheral areas of the city.

To understand Bogotá's transit needs, one must first understand its geographic layout. In Bogotá, income strata are drastically segregated geographically. Higher-income areas are more centrally located, while the poorest areas are located in the city's periphery, mostly in the southern and western edges of the city. Additionally, only a small percentage of this densely populated city's inhabitants have a car, so the majority of inhabitants depend on public transportation, walking, or biking (Cain et al. 2006). Before the implementation of TransMilenio, urban bus routes were sold to and provided by private transportation companies. The city oversold its permits, and heavy competition between route owners led to low-quality service, high congestion, severe pollution, and frequent accidents (Turner, Kooshian, and Winkelman 2012).

The long preferred solution to the city's transportation problem was always a subway or metro system. Initially, Mayor Peñalosa (who held office from 1998 to 2000 and was reelected in 2016) had plans to build an integrated transit system composed of a metro and a network of bus lines. However, this plan was too expensive. Focusing on a comprehensive bus network alone would provide for much more coverage at a lower cost than a metro could (Bonilla-Chacín 2013). His plan was to revitalize and stimulate growth through an improved transportation system. The system that was eventually approved by city officials, the TransMilenio, was based on Curitiba Brazil's 20-year proven bus system (Bonilla-Chacín 2013). Private transportation franchisees, who had considerable economic and political influence in the city at the time, initially opposed the project but became supporters once they were allowed to participate in the preliminary bidding process.

Plan Implementation

The TransMilenio system is a public-private partnership. The Institute of Urban Development (IDU) provides the infrastructure (which includes the grid system, platforms, and stations); seven private firms, chosen through a public bidding process, operate the bus system and are paid per kilometer covered; two other private firms collect fares (Bonilla-Chacín 2013). The whole TransMilenio system is managed, planned, and monitored by TransMilenio SA, a public company that reports to the mayor's office (Wright 2014). The majority of revenues from TransMilenio ticket sales are distributed to the private bus operators, divided between trunk line and feeder services (Wright 2014). Tickets are paid for through the use of smart card technologies, and the fleet of buses is monitored through GPS to optimize service (Cain et al. 2006).

Initial plans for the TransMilenio were ambitious. In 1998, Mayor Peñalosa planned a bus network that would cover 85 percent of the city's daily transit demands and would run through 241 miles of the city's main avenues (Cain et al. 2006). The TransMilenio master plan included six phases to be implemented over several decades (Montezuma 2005). Estimated total costs for the project, including vehicle costs and fare collection, would be around US$3.32 billion (Bonilla-Chacín 2013). The implementation of phase I was highly successful. By the end of Peñalosa's 3-year term, 25.6 miles were constructed. Phase I of the TransMilenio demonstrated that a bus mass transit system with high capacity and high quality could be built in a timely manner (Cracknell 2003). The Trans-Milenio also measurably reduced private vehicle usage by 10 percent. A metro or subway would have only accounted for 18 miles of coverage and would have been 10 times the cost of the TransMilenio system (ITDP 2007). Phase I was estimated to cost US$240 million (US$5.9 million per square kilometer). Financing came from four separate sources: 46 percent from the gasoline tax, 20 percent from the national budget, 6 percent from the World Bank, and 28 percent from local funds (Bonilla-Chacín 2013). Phase I of the TransMilenio bus system was carrying 792,000 passengers daily by 2003 (Bonilla-Chacín 2013).

But phase II of the TransMilenio expansion was not as successful. This phase was much more expensive than the previous phase and was delayed by several years. The three corridors of phase II were estimated to cost US$545 million at an average of US$13.3 million per square kilometer (Bonilla-Chacín 2013). The higher costs of phase II were due in large part to the larger area of land that had to be purchased: while only 600 new plots had to be purchased for phase I, the city had to purchase 4,000 new plots for phase II (ITDP 2007). Funding for phase II has also been increasingly sourced from the national government (at 66 percent) and the rest coming from a local fuel surcharge (Cain et al. 2006). Construction of phase III of the program was also delayed significantly.

Nevertheless, by 2014 the system had 115.5 kilometers of route with 131 stations, and almost two million passengers used the system daily. Approximately 30 percent of Bogotá's public transport demand was met through this system (TransMilenio 2014). Current development of the TransMilenio is slower than planned, and there is much uncertainty as to where the money for future phases will come from (Hutchinson 2011). These uncertainties revived the public debate over the merits of a bus network over a rail system.

Challenges

The TransMilenio system presents some challenges. One is information: there are no maps of the bus network outside of the stations or on buses, and the few maps that are inside the stations are clustered in one area even though stations are usually around three city blocks long (Hutchinson 2011). Another concern is overcrowding. Most of the trunk-line buses fit a maximum of 160 people, and during rush hour these buses are almost always packed to full capacity. Passengers often cannot get on the first bus and must wait for subsequent buses to get on. The price of the Trans-Milenio is also considered high by many of its lower-income passengers. Among non-users a common complaint is the increased congestion in Bogotá's streets because of constant TransMilenio construction along the city's main avenues (Batarce et al. 2015).

TransMilenio allows its riders to rapidly reach jobs far from their homes. The building of the TransMilenio infrastructure also allowed for increased access to bike paths as well as improved pathways and bridges for pedestrians. Road safety has significantly increased and noise pollution has been reduced. Recently, the municipality has introduced SITP (Integrated Public Transport System) in order to integrate all bus usage into the TransMilenio system. Additionally, many other cities around the world have been inspired by the TransMilenio and have implemented similar BRT systems.

LIGHT RAIL IN JERUSALEM

Background: Jerusalem's Economy, Culture, and Context

Jerusalem, the capital city of Israel and one of the oldest cities in the world, is located between the Mediterranean Sea and the Dead Sea. It has a population of 839,823 and covers about 48 square miles (United Nations Data 2016). It is the home of several religious landmarks, including the Western Wall, a sacred site of prayer in Judaism; the Resurrection Rotunda, which is a Christian landmark; and the Dome of the Rock, a seventh-century Islamic shrine (Karsten 2015). Jerusalem has a diverse population, with approximately 500,000 Jewish residents and 300,000 Arab residents (Rudoren 2014).

As the capital of Israel, the city is a focal point of the Israeli-Palestinian conflict. Jerusalem also faces economic issues. Jerusalem's economy is based mostly in the service industry, such as government, education, and tourism. Partially because of orthodox religious affiliation, many households in

Jerusalem are single-income. There is also a higher percentage of citizens living in poverty in Jerusalem when compared to other Israeli cities such as Tel Aviv and Haifa (Advameg City Data 2008). The increasing number of tourists and pilgrims stimulates the local economy, and individual income for citizens has risen steadily over the past 45 years, although extreme poverty is still concentrated among sections of the Muslim population, particularly in the Old City, and among Orthodox Jews and Jews from Africa and Asia (Wasserstein, Perowne, and Prawer 2016).

The Jerusalem Light Rail

The idea for a public transit system in Jerusalem emerged as a solution for the congestion in the city center. As the city grew throughout the 1980s and 1990s, the residential areas expanded. Between 1988 and 1994, the city population increased by 15.6 percent with the majority of incoming residents moving to places outside of the city center. The use of public transit decreased from 60 percent in the 1980s to 39 percent in 1996. Congestion presented an aesthetic and environmental issue, with heavy traffic in Jerusalem's center producing air pollution, noise, and crowds that deterred visitors (Daniel and Render 2003). By the end of the 1990s, drivers in Jerusalem were frustrated with constant delays, and the city realized it had to reevaluate its policies for investment in road infrastructure rather than improvement of the public transit system. The Ministry of Transportation was inspired by Tel Aviv's successful suburban rail system and by a growing environmental movement that rallied for a rail investment, citing benefits like reduction in noise and air pollution and mitigation of the excessive land use of urban sprawl by making the city center more accessible (Daniel and Render 2003).

An extensive plan was drawn up, including the establishment of a "ring road," an outer and inner path encircling the city, which would couple roadways with the light rail system. This ring road was to include Jaffa Road, a major road in Jerusalem that already serves as a crucial urban corridor for buses. When assessing the investment, the city projected that establishing a rail system coupled with the preexisting bus system would decrease pressure on and saturation of the buses and roads, decreasing congestion and pollution (Daniel and Render 2003).

Approval and construction of the light rail system began in 2002 and was completed in August 2011 with the operation of the Red Line part of the system (Mota 2015). The Red Line runs from northern Heil Al Avir, along the east side of the city, down to the Old City area, and then moves west along Jaffa Road and continues down to Mount Herzl. While the Red Line is currently the only operational line, a new Blue Line has been approved and was announced in January 2016, with plans for a Green Line to be developed in the near future (*Times of Israel* Staff 2016).

CityPass Group, a consortium of companies from around the world, was tasked with developing, building, and operating the new line for 30 years. The total budget for the project has grown to about €925 million (US$1.04 billion; Mota 2015). In 2013, the Jerusalem Regional Planning Committee approved an allocation from the Ministry of Transportation's budget to extend the light rail line (*Railway Gazette* 2013). In 2015, a multinational infrastructure company, Veolia, sold its 5 percent stake in City-Pass in response to boycott, divestment, and sanctions by activists who were making it difficult for Veolia to sign public contracts in Europe, the United States, and the Middle East. Veolia ultimately decided to separate from its involvement in Israel (Abunimah 2015).

Benefits and Criticisms

The benefits of the Jerusalem Light Rail system are primarily economic and environmental. It has renewed the city center by encouraging new businesses and housing developments along Jaffa Road, rejuvenating an area that had been deteriorating in the 1990s (*Times of Israel* Staff 2016). A study of air pollution in Davidka Square (located in the midway section of the Red Line) by the Jerusalem Municipality Environmental Department indicates that as traffic was slowly eliminated for the installation of the light rail, between 2002 when construction began until completion in 2011, annual average levels of nitrogen dioxide and nitric oxide declined by 80 percent, from a high of 151 parts per billion (ppb) in 2002 to a low of 32 ppb in 2011 (Frieling and Levy n.d.). After its third year of operation, Israel's Ministry of Transportation and the Jerusalem Municipality won the Light Rail Award 2015 in the Best Environmental & Sustainability Initiative category for cutting air pollution in Jerusalem by 85 percent

(Ahronot 2015). In addition, the light rail transported about 140,000 passengers a day in 2015 and is a viable and preferred public transit option, helping unite many communities in Jerusalem (Mota 2015).

However, the Jerusalem Light Rail system has incurred criticism and been subjected to sabotage and violence. It was criticized for being a financially wasteful prestige project, originally estimated at €120 million (about US$135.5 million) but eventually reaching a budget of €925 million (US$1.04 billion; Mota 2015). City officials argued that the light rail would balance the exhausted bus system and mitigate congestion, while providing a more direct route less susceptible to traffic (Daniel and Render 2003). There were also concerns that the light rail would become a target for terror attacks (Nolte 2016).

Conclusion

According to Jerusalem Mayor Nir Barket, "With the construction of a local bus network, along with a high-speed train from the center of the country, the city's residents will be able to enjoy shortened travel times, less traffic, cleaner air and exciting urban renewal" (*Times of Israel* Staff 2016). The Jerusalem Light Rail has been praised as an artery of coexistence that serves both Arab and Jewish neighborhoods. But it has also been a location where the conflict between Israelis and Palestinians has been expressed. Nevertheless, the light rail has proven beneficial to Jerusalem's economy and environment and has made Jerusalem a more sustainable city. Its positive contribution to air pollution control and congestion along with its beautiful design are obvious benefits. The rail line between Jerusalem and Tel Aviv will do even more to promote sustainable travel between these two cities.

HIGH-SPEED RAIL IN CHINA

Understanding the Problem

Over the past decade, China has undertaken a massive expansion of its high-speed rail networks. Today, China has the world's longest high-speed rail network with more than 19,000 kilometers of track in service, more than the rest of the world's high-speed tracks combined (Minter 2016).

The network extends to 28 of China's 33 provinces, connecting distant regions and opening up previously unimaginable travel across the country. Since high-speed rail service in China was introduced in 2007, daily ridership has grown from 237,000 to 2.49 million. Annual ridership reached more than 1.1 billion in 2015, making the Chinese high-speed rail network the most heavily used in the world (Jin 2012). In the next 5 years, China aims to add 30,000 kilometers of high-speed rail tracks, more than doubling the current network (JLL Staff Reporter 2016). By 2020, the high-speed rail network will connect all provincial capitals and cities with a population of more than 500,000—serving around 90 percent of China's population (Stokols 2015).

State planning for China's high-speed rail network began in the early 1990s to replace an obsolete railway system that prevented balanced development among different provinces (Boquet 2011). At the time, the average speed on Chinese railways was below 60 kilometers per hour due to outdated technologies and overcrowding, and commercial train service was steadily losing market share to airline and highway travel (Zheng and Kahn 2013). A series of campaigns led by the Ministry of Railways (MOR) in the late 1990s and early 2000s modernized and upgraded existing rail lines, but the highest speed did not exceed 150 kilometers per hour. The first set of high-speed bullet trains opened in 2007, and the next year China unveiled the Beijing-Tianjin Intercity Railway in time for the 2008 Summer Olympics. The Beijing-Tianjin line between northern China's two largest cities was the first in the country to accommodate commercial trains with a top speed of 350 kilometers per hour (217 miles per hour; Zheng and Kahn 2013). By 2011, China had the world's longest high-speed rail network.

The opening of the Beijing-Shanghai line in July 2011 was a major achievement for the Chinese government. However, a fatal high-speed railway accident near Wenzhou a month later created significant safety concerns, slowing the pace of expansion. In addition, low ridership, cost, and environmental issues cast doubts on the high-speed rail system (Jin 2012). Despite these setbacks, by early 2012 the Chinese government renewed investments in high-speed rail to rejuvenate the slowing economy. Over the course of the year, the budget of the Ministry of Railways grew from $65 billion to almost $100 billion (Rabinovitch 2013).

China continues to move forward ambitiously with high-speed rail development; the country's most recent Five-Year Plan (2016–2020) plans to invest 3.5 trillion yuan (about US$508 billion) into new railway construction (Minter 2016).

The system's rapid expansion hasn't been without problems and controversy. Doubts about the continued investment in high-speed rail development are rising, largely due to concerns about China's growing debt and the low ridership on many of the rail lines.

Improving Access to Cities

China's high-speed rail has transformed travel within the country. As of 2014, high-speed bullet trains carried twice as many passengers each month as the country's domestic airlines and had an annual growth rate of 28 percent (Bradsher 2013). Train ridership has soared partly because China has set fares on high-speed rail lines at a little less than half of comparable airfares. That has resulted in many workers, as well as business executives, switching to high-speed trains (Bradsher 2013). As the high-speed rail dramatically cuts traveling times between cities, the increased mobility of commuters is key to driving widespread growth, especially as China lays foundations for sustainable growth in its rapidly growing cities. The high-speed rail connection between Beijing and Shanghai, for instance, more than halved the 1,318-kilometer (819-mile) journey time from 12 to 5 hours (JLL Staff Reporter 2016). Factory workers in cities can now travel home to their families more than once a year. Small business owners and manufacturers can now afford to travel more frequently to distant commercial hubs, gaining greater access to client bases. A more mobile workforce and newly accessible markets may narrow the country's geographic and economic disparities.

China's extensive high-speed rail program is reshaping the location of China's industries, and even services, to open up new business clusters and growth opportunities (Zheng and Kahn 2013). Working as an economic corridor, it optimizes regional economic development and supports increased resource circulation. For example, after the Beijing-Tianjin Intercity Railway began operating in 2008, both cities saw a boost in economic activity as trade volume increased and new investment and business opportunities emerged. In the first 2 years after the rail line opened, Tianjin experienced economic growth of 16.5 percent (Ying 2014).

Improving transport infrastructure between nearby cities also offers one strategy for mitigating the "megacity quality of life challenge"; that is, traffic congestion, pollution, overcrowding, and so forth (Zheng and Kahn 2013). Because high-speed rail effectively brings cities closer together by reducing travel times, it allows people to enjoy many of the benefits of big cities while living in "second-tier" cities far from the pollution and congestion. For example, cities such as Tianjin, Nanjing, and Shaoguan are located between 100 and 750 kilometers (62 and 466 miles) from China's megacities of Beijing, Shanghai, and Guangzhou. By making second-tier cities attractive to those who would otherwise flock to global hubs, bullet trains could act as a safety valve for crowded megacities in the developing world and ease the effects of overpopulation. Workers and firms who require infrequent face-to-face meetings with firms and government officials will also enjoy a reduction in rents by relocating away from the megacities (Zheng and Kahn 2013).

Rising Debt

Some have begun to question the success of China's high-speed rail. Critics have concerns about safety, high ticket prices, low ridership, financial sustainability, and environmental impact. Despite claims of affordability, high-speed rail prices on many lines are too high for working-class citizens to afford. A World Bank survey found that the average income of high-speed train passengers was 35 to 50 percent higher than that of conventional train passengers (Bullock et al. 2014, 12). Expensive luxury services, like the sleeper service offered between Shanghai and Chengdu that costs almost $350, have had little commercial success (JLL Staff Reporter 2016). However, critics are mainly concerned about China's rising public debt.

The rapid expansion of high-speed rail in China is the result of massive investment from the Chinese government's economic stimulus package and the low cost of building rail in China. Several factors contribute to these low costs of construction, including China's low labor costs and the government's ability to easily procure land. A 2014 World Bank analysis estimated that China spends between $17 million and $21 million per kilometer on high-speed rail, compared to $25 million to $39 million in Europe, and as high as $56 million in California (World Bank 2014a). Even with these advantages, however, the costs have been considerable. In May 2016, state-owned China Railway Corporation, the operator of China's rail network, reported that its debt had grown 10.4 percent in the past year and now exceeded $600 billion; in 2014, roughly two-thirds of that debt was related to high-speed rail construction (Minter 2016). And costs are predicted to rise further. China's most profitable line is the well-traveled Beijing-Shanghai corridor (Minter 2016). Now that most heavily trafficked areas are served by high-speed lines, construction is expanding to China's less-populated and less-developed western regions, building lines over greater distances and across more difficult terrain, in part as a de facto fiscal stimulus. While some train construction may not appear successful initially, new rail lines encourage development, and increased ridership will then follow.

Most of the rail construction was financed with bank loans that must be rolled over as often as once a year. While using short-term loans made the financing look less risky on the balance sheets and held down borrowing costs, the reliance on short-term credit has left the system vulnerable to any increase in interest rates (Bradsher 2013). As far back as 2010, prominent voices in China had warned that spending on high-speed rail could lead to a debt crisis and that the same benefits could be achieved with conventionally built lines that cost about one-third as much. Traditionally ignored, concerns about rail-related debt are now gaining traction, leading to prominent calls to break up the massive China Railway Corporation (Minter 2016). So far, however, the government has yet to slow down any high-speed rail development.

The backlash of China's rising debt is being felt most acutely abroad, where China has hoped to export its technology. Those ambitions have run into major difficulties, as many of the countries to which China had hoped to sell high-speed technology are now scaling back their plans "due to huge building and operating costs" (Minter 2016). Thailand has opted to shorten a planned, Chinese-built high-speed rail line over financing questions. Indonesia agreed to another Chinese project only after China agreed to build the line without Indonesian government money or loan guarantees. Mexico cancelled a Chinese high-speed rail project outright, ultimately citing budget constraints.

Conclusion

China's unique political structure has allowed it to implement high-speed rail quickly and efficiently across the country (Garmendia, Ribalaygua, and Ureña 2012). However, the massive investment in high-speed rail has led to a rapid escalation of state debt. High-speed rail has obvious benefits, but the question for China now is how they will sustain growth in this sector while managing their rapidly growing debt. Understanding and addressing passenger needs are critical to achieving the full impact of the high-speed rail network. With sufficient population density, nearby secondary cities, and congested traffic on competing modes such as highways, high-speed rail in China can be cost-effective while providing wider economic benefits.

TRANSIT IN THE UNITED STATES

While we've examined public transportation options in Colombia, Israel, and China, in the United States any discussion of transit must include personal transportation because of our suburban sprawl style of land use. A sustainable transportation system would also include mass transit that is based on renewable energy, helping to reduce greenhouse gas emissions and to reduce local air pollution.

Electric Vehicles

Electric Vehicles (EVs), which include both battery electric vehicles (BEVs) and plug-in hybrid electric vehicles (PHEVs), are gaining traction worldwide. According to the International Energy Agency (2016, 8), electric vehicles "have the capacity to increase energy efficiency, diversify transport energy carriers, and reduce their carbon intensity, supporting the integration of variable renewable energy in the power generation mix and transferring to the transport sector GHG emissions mitigations occurring in power generation. BEVs and PHEVs are also well equipped to reduce emissions of local pollutants in high-exposure areas such as urban environments, where they would also reduce noise levels." For commuters who can charge at home, there is no trip to the gas station, and a full charge costs only a few dollars. Electric cars are also surprisingly powerful. And overall, electric cars produce roughly 40 percent less carbon dioxide and ozone than conventional cars, even when factoring in emissions from power plants that produce the electricity to power the cars and manufacturing and disposal (Bullis 2013).

However, there are many barriers to widespread electric vehicle adoption. First, electric vehicles are typically more expensive up front than traditional cars, which can be a significant deterrent for consumers, even if they are cheaper to operate in the long run. In addition, there are issues with range limitation (how far the car can drive on one charge), causing drivers to have "range anxiety" and requiring advanced planning for refueling, which is not convenient and certainly not always easy. Consider that in the United States, there are about 10,000 publicly available electric vehicle chargers compared with about 114,000 gas stations (Patterson 2015). But these cost and range issues are largely dependent on battery technology: a more powerful and efficient battery could allow for longer ranges and could also reduce the overall price of electric cars.

Another significant barrier to widespread adoption is lack of charging infrastructure to support a growing network. This is often thought of as a "chicken and egg problem"—what comes first, the electric vehicles or the infrastructure to support their deployment? Charging infrastructure refers to all the hardware and software that ensures that energy gets from the grid to the vehicle. The majority of people who use electric vehicles currently charge their cars at home. But in urban areas like New York City, where multifamily homes are common and many people park on the street, home charging is often not possible. These urban dwellers must rely on public charging stations, where you have to pay to both park and charge. Ryan Bradley wrote in *MIT Technology Review*: "The majority of people who own electric cars today charge their vehicles at night in the garage or driveway of a home that they own. But that is changing fast. The vast majority of new electric car sales are to people who live in cities, which makes sense because the vehicles are very efficient at stop-and-go urban driving. Charging at home at night can be impossible for people who rent an apartment or lack a garage" (Bradley 2016).

Expanding the number of public charging stations could help expand electric vehicle deployment. One study found that a 10 percent increase in the number of charging stations per million people in a city would result in a 10.8 percent increase in the market share of electric vehicles in that city (Mixon 2015). However, it is not as simple as having more charging stations. The network also needs to be convenient and user-friendly. There are currently 1,855 fast-charging stations in the United States, but German and American automakers use different connection standards than Asian manufacturers, and Tesla's connections are different still (Wittenberg 2016). The network of different charging stations is confusing and inconvenient. Some progress is under way—Nissan and BMW recently announced a partnership to install fast chargers that will be equipped with two of the most popular connectors (CHAdeMO and SAE), while Tesla developed an adapter for use with CHAdeMO plugs. But the availability and ease of use of these charging stations will certainly play an important role in electric vehicle use and adoption. In addition, the public's electric vehicle drivers have reported their experience is mixed—drivers say that even when they can find an open station, it is often broken due to software or hardware problems. There have also been reports of poor etiquette, with people unplugging cars that aren't their own or simply using the spots for parking instead of charging.

Tesla Motors Inc.

One car company—Tesla Motors Inc.—has started to change the electric vehicle game by making an electric vehicle with batteries that recharge quicker and have a lower cost than those of competitors. Tesla's choice of small lithium-ion batteries—the kind designed for laptops—allows them to save on manufacturing costs compared to the costs of battery types in other electric cars (Bullis 2013).

Tesla was founded in 2003 by Martin Everhard and Marc Tarpenning. Elon Musk—a PayPal cofounder—became the company's chairman of the board in 2004 and CEO in 2008 and has invested heavily in the company. The founding of Tesla was based on a vision that batteries were getting better and more efficient, and if the trajectory stayed the same, Everhard and Tarpenning believed they could build a vastly superior electric car. They figured it wouldn't be too difficult considering the segmented nature of the auto industry; everything from styling to windshields to electronics was outsourced to other companies (Baer 2014).

Tesla's flagship car, the Roadster, was unveiled in 2008 at a cost of $109,000. The Tesla Model S followed in 2012 at $57,400 and the Tesla Model X in 2015 at $80,000. The U.S. Department of Energy granted Tesla a $465 million loan in 2009 to support the commercial-scale development of advanced auto technologies, and Tesla paid off the loan in May 2013, about 9 years early (Kumparak, Burns, and Escher 2015). It was the first American car company to go public since Ford in 1956. Tesla's newest model, the Tesla Model 3, is a five-passenger sedan with a 215-mile range and a starting price of $35,000—its cheapest model yet (Boylan 2016). In late 2016, there were about 400,000 preorders for the car, expected to ship about the same time this book is published in 2017. Tesla also began building a network of electric car charging stations throughout California in 2012; there are currently 274 stations in North America and 613 world-wide, and their superchargers can fill up a Tesla in 30 to 40 minutes, much faster than the typical 3 to 4 hours at public charging stations (Bradley 2016). A Morgan Stanley auto analyst called Tesla the "world's most important car company," and a 2014 survey found that Tesla's Model S is the "most loved vehicle in America" (Baer 2014). The huge demand for the newest Tesla model is indicative of the rising demand for electric vehicles. But the growth of Tesla was not without issues. Manufacturing setbacks have caused each model of the Tesla to be delayed. In 2009, Tesla also had to recall 75 percent of its Roadsters because of issues with loosened bolts. And in 2016, a Tesla Model S with an activated autopilot system was involved in a fatal crash.

Most recently, in June 2016 Tesla announced a $2.86 billion plan to combine with SolarCity Corp.—the largest player in the U.S. residential solar market, whose chairman is Elon Musk—to create the world's first vertically integrated sustainable energy company. Energy generation, storage, solar panels, home batteries, and the electric vehicle would all be offered by one single, trusted brand. When this deal was first announced, Tesla shares dropped more than 10 percent. One analyst at

Morgan Stanley warned that expanding into a non-auto business exposes Tesla to "untested cost, competitive and regulatory forces" (Stewart 2016). Critics have accused Musk of using Tesla shares to bail out a struggling SolarCity (Ramsey and Sweet 2016). However, Musk has stood behind this decision, arguing that this is part of his vision to accelerate sustainable energy. According to Musk's master plan, the idea is to "create a smoothly integrated and beautiful solar-roof-with-battery product that just works, empowering the individual as their own utility, and then scale that throughout the world. . . . We can't do this well if Tesla and SolarCity are different companies, which is why we need to combine and break down the barriers inherent to being separate companies" (Stewart 2016). Tesla and Musk aim to not just change the electric vehicle game, but the entire energy system.

Conclusion

Electric vehicles offer a promising solution to a sustainable form of personal transportation. With the infrastructure to support them, they can be especially effective in cities, where drivers take shorter trips. The electric vehicle market is primarily dependent on (1) technology and (2) government incentives, both of which can make electric vehicles cheaper and more convenient than traditional petroleum-powered cars. Bloomberg New Energy Finance estimates that electric vehicles will be a cheaper option than gasoline or diesel cars by the 2020s. They predict sales of electric vehicles will hit 41 million by 2040, which is 35 percent of new light-duty vehicle sales. This forecast work is largely based on the drop in battery prices. Lithium-ion battery prices have dropped 65 percent between 2010 and 2016 and will continue to go down in price (MacDonald 2016).

Some efforts have been made to incentivize charging infrastructure. In July 2016, the White House announced a series of actions to support electric vehicle charging infrastructure, including $4.5 billion in loan guarantees to support commercial-scale deployment of EV charging facilities; publishing a guide to electric vehicle funding, financing, and technical assistance; and launching the FAST Act process to identify zero emission and alternative fuel corridors (The White House 2016). But it is no surprise that the places where we are starting to see success with electric vehicles are in cities. According to the International Council on Clean Transportation: "Urban areas, with their concentrations of vehicle ownership, distinctive driving patterns, and particular transportation needs, are critical hubs for development of that market—especially now, while it remains in its nascent stage" (ICCT 2015, 1). The cities of San Francisco, Los Angeles, New York, Seattle, and Atlanta contain more than half of all electric vehicles in the United States (Lavrinc 2013). In fact in San Francisco, because of federal, state, and local consumer incentives, the battery electric vehicle is lower in cost per mile than the average gasoline-powered car (ICCT 2015, 3). And according to the International Energy Agency (2016, 6): "EVs of all types lie at the heart of future sustainable transport systems, alongside the optimization of urban structures to reduce trip distances and shift mobility towards public transportation." It is unlikely that this degree of federal support will survive the Trump administration, although it is possible if charging infrastructure is prioritized by state and local governments and local businesses.

ANALYSIS OF MASS AND PERSONAL TRANSIT: BUS RAPID TRANSIT IN BOGOTÁ, LIGHT RAIL IN JERUSALEM, HIGH-SPEED RAIL IN CHINA, AND TESLA IN THE UNITED STATES

Transit as an Issue of Values

As I indicated in chapter 2, one of the main functions of cities is to bring people in contact with each other. This includes work, culture, entertainment, recreation, socializing, and a wide variety of activities. How one moves around an area and how diverse the stimulation wanted is a fundamental preference of city dwellers. Preferences can include comfort, speed, status, cost, and a variety of tangible and intangible factors. In each of the cases presented here, we see the difficulty of changing people's transportation habits, but we also see that new methods of transit happen all the time. People want to move around and are always looking for new and more effective modes of transport. People are open to the new technologies described here, and people keep trying different ways of getting around.

The mode of transport we use is related to where we live and where we work, which is also in part a reflection of our values. If we prefer a rural or suburban environment to live or work in, then we will require personal transit. If we are more interested in cities, we leave open the possibility of walking or using mass transit.

Transit as a Political Issue

While personal transport (cars, bikes, and walking) involve private property, the reliance on public roads, sidewalks, ports, trains, and airports make transportation inherently political. Some folks lobby for transportation infrastructure such as train stations, roads, and bridges. Some protest against siting these facilities near their homes, seeking to preserve the character of the neighborhoods they now enjoy. Companies and localities lobby for the funds and employment gained by building and operating transit facilities.

The politics of transit can become intense and high in conflict. In New York City, former Mayor Michael Bloomberg faced intense opposition when he sought a congestion charge on motor vehicles in New York's central business district. He lost that battle. Over half a century earlier, New York's master builder Robert Moses built the Cross Bronx Expressway before locals knew what was happening but then failed to build an expressway that would have cut Greenwich Village in two. Jane Jacobs's classic work *The Death and Life of Great American Cities* was informed by this battle and began a major movement questioning the impact of transportation infrastructure on neighborhood life.

Transit as an Issue of Science and Technology

Ever since we learned how to tame animals, build wheels, and build engines, we have applied technologies to moving ourselves and our material goods around. Today, our transit technology is dependent on fossil fuels, but the technology of electric transit is advancing, leaving open the possibility of a renewable energy–based transit system. Other advances in high-speed trains, driverless cars, and computer-controlled subways open up possibilities of improved speed and comfort of transit.

The value of movement and the difficult politics of siting can lead to new demands for transportation technology. Wedding information technology to transport opens up the possibility of more efficient fare-charging systems, off-peak pricing, and more responsive scheduling of mass transit. Improvements in speed, comfort, and amenities such as WiFi make the process of moving around more comfortable and convenient and can be expected to continue.

Transit as an Issue of Public Policy Design

Some transit promotes socially desirable goals. Bus service is provided to help families send their children to school. In New York City, the mass transit system is used to avoid the need for a separate transit system for some schoolchildren. Bridge tolls can be used to subsidize mass transit, making driving more expensive and riding the subways less expensive. The construction of the Interstate Highway System possibly inadvertently led to a move away from cities and subsidized the settlement of America's suburbs.

Investment in one type of transit facility over another and locating a transportation facility in one place rather than another has a profound influence on patterns of land use, ecosystem well-being, and even community cohesion. The type and amount of subsidies and investments made in transportation have a profound ripple effect throughout political, social, and economic life and are among the most consequential decisions made by governments.

Transit as an Issue of Management and Organizational Capacity

The complexity of moving billions of people each day from place to place should not be underestimated. Air traffic control, train scheduling, operation and maintenance of mass and personal transport, and planning, constructing, and maintaining transportation infrastructure are enormously complex sets of activities. The number of accidents and deaths taking place each day provide a small indication of the extent and difficulty of these tasks. To a considerable extent these capacities must adapt rapidly to new technologies and growing demand. As economic development takes place, as rural folks move to cities, and as societies evolve, the public's demand to move from place to place increases.

Management innovations are driven by the needs of transport, and these trends can be expected to continue. Colombia's new bus system required a more coordinated and informed organizational capacity than its "free market" predecessor of small unscheduled private buses. Simply getting in a bus and driving it where customers might be was replaced by a scheduled, synchronized, more sophisticated system. This required more highly trained and coordinated staff to implement. Without this capacity, the creative design and new technologies would not have worked.

Transit as a Multidimensional Issue

When we examine the cases in sustainable urban transportation presented here, it is clear that politics, management, technology, and subsidization are all involved in developing and maintaining a transport system that is less resource intensive. For personal transit, the technical issue of battery range and the speed of recharging will tend to dominate. For mass transit, the politics of siting and subsidization will be most important. Organizational capacity will be needed in all cases. The examples presented here are intended to convey confidence that change is possible, but also that transition to a renewable resource–based transport system will be complex and will include failures as well as successes.

8

THE BUILDING OF THE SMART GRID

———

Cases of Microgrid Development

Energy in the sustainable city must be based on renewable sources and must include newer technology that allows for more efficient and resilient energy systems. Traditionally, energy is generated at a central location and then distributed out to where it is needed. A microgrid is a localized energy generation system that can disconnect from the larger grid and function independently. A microgrid functions at the same voltage as the larger grid but can be disconnected to function on its own in case of disruption and power a small section of the grid, referred to as "island mode." There are many benefits to using a microgrid. Microgrids enable small communities to be more energy independent and allow for the opportunity to cut costs and use energy sources that are renewable, more energy efficient, and better for the environment.

As microgrid technology develops, it presents a solution for areas that lack the infrastructure to meet energy needs, such as in developing countries. It also provides recovery and resilience options for regions that are vulnerable to extreme weather events. However, the transition from the current structure of centralized energy generation to localized, independent energy grids requires greater levels of investment, research, and development. Governments and communities must work with private investors, contractors, project managers, and energy companies to develop strategic energy plans that are region specific. This case study looks at three examples of microgrid implementation: the first at New York University in New York City, the second in several African countries through the Power Africa initiative, and the third in the city of Higashimatsushima, Japan.

NEW YORK UNIVERSITY

New York University (NYU) in New York City is one of the largest private universities in the United States (NYU 2016a). In March 2010, NYU released its Climate Action Plan, a plan for improving the university's sustainability practices that details the university's greenhouse gas emission reduction strategies and its goal to achieve climate neutrality by 2040 (NYU 2010). The university also set conservation goals to be met through individual behavioral changes and upgraded buildings. The latter requires operational innovations to run buildings more effectively, such as enhanced building management systems, advanced controls, active monitoring, and operator training (NYU 2016b).

Transitioning to Combined Heat and Power

New York University has historically been committed to distributed energy generation (on-site or localized generation), with connected campus facilities, high reliability, and the pursuit of cost-effective, cutting-edge, energy-efficient technologies (Griffin and Merrihue 2013). NYU has produced power on site since 1960, when it installed an oil-powered cogeneration plant. Since 2011, when the university transitioned from oil-powered to natural gas–powered technology, it has utilized a new combined heat and power (CHP) facility, which combines electricity generation and useful heating processes in a thermodynamically efficient way (Berkeley Lab 2016).

The transition to a natural gas–based combined heat and power system was driven by state air pollution regulations; the old system had to close by 2008 because its 1970s-era diesel-engine construction was incapable of meeting new air quality standards. In conjunction with this closure, NYU joined New York City Mayor Michael Bloomberg's Carbon Challenge, which encouraged universities to reduce greenhouse gas emissions 30 percent by 2017 from 2006 levels (Overton 2014).

New York University also wanted to support more of its own power needs. Its oil-powered plant was only able to provide power to seven out of NYU's 70 core buildings, and only when the system was in island mode (Overton 2014). One of the benefits of a smaller grid is that it can continue to function in times of disruption. A self-contained grid can act as an isolated entity and, in the case of universities, keep students safe and allow power to continue to flow for research purposes, even in the event of a large-scale power outage (Pyper 2013).

New York University's newer system is a combined heat and power facility that has an output capacity of 13.4 megawatts, twice that of the old plant's capacity, and consists of two 5.5-megawatt gas turbines for producing electricity, coupled with heat recovery steam generators and a 2.4-megawatt steam turbine (Berkeley Lab 2016). The upfront cost of the upgrade was roughly $126 million, made possible through tax-exempt bonds arranged through the Dormitory Authority of the State of New York and through NYU tuition and fees. Though the university is con-nected to the Con Edison grid in order to purchase energy when demand exceeds on-site capacity, the NYU microgrid is able to island from the distribution grid (Berkeley Lab 2016).

The CHP system, also known as a cogeneration plant, produces both electricity and steam, and instead of wasting that steam uses it for dis-trict heating, improving the overall efficiency of the system (Reegle 2016). This dramatically reduces wasted energy and thus enabled NYU to reduce greenhouse gas emissions by more than 30 percent. In addition, the entire combined heat and power system at NYU can be controlled by a high-tech computer system rather than the manual valve-adjustment controls of the oil-powered system (NYU 2016b).

Because NYU is located in lower Manhattan, a densely populated area that lacks open space, the cogeneration plant is located underground near Washington Square Park. Construction began in September 2008, with the initial excavation in front of NYU's Courant Institute of Mathematical Sciences. According to Thomas Overton (2014) in *Power Magazine*:

> Because of the site limitations, the entire system had to be squeezed into a 200-foot-long vault that was only 60 feet deep and 40 feet wide. Rigging the gas turbines, heat-recovery steam generators (HRSGs), and other heavy equipment into the new plant was a challenge in the narrow, mostly residential streets of Greenwich Village, but for the experienced construction crew used to working in the tight confines of Manhattan, it was largely business as usual. The project was completed in January 2011 after 28 months of construction.

As an added bonus, the removal of the old plaza to install the plant provided an opportunity for the construction of a new community park in its place. The old sidewalks of the plaza were replaced with seating and gardens outside of the Courant Institute.

Microgrids in Action

This update served the university in an emergency situation in 2012. In October 2012, Hurricane Sandy swept the eastern coast of the United States, severely affecting the coasts of New Jersey and New York. It was the second costliest hurricane in United States history after Hurricane Katrina, with damage in the United States amounting to $71.4 billion (AOML 2014). The hurricane caused massive flooding, killed 48 people, and injured and displaced many more. The city's energy grid was also damaged after a massive storm surge hit Con Edison's substation at East 14th Street in Manhattan, and the transformers at the station exploded. This caused outages all across Con Edison's grid, leaving most of lower Manhattan without power, including NYU, which was connected to Con Edison's grid at the time. When the outages caused a pull of electricity from the NYU grid, the CHP system began to push an excessive amount of power out to the larger grid, resulting in the NYU system going into island mode and disconnecting from Con Edison's grid. This meant that while much of lower Manhattan was plunged into darkness, NYU's campus lights stayed on.

The powered part of campus became a safe haven for students in the coming days as power had yet to be restored to Manhattan, and in the coming weeks the campus became a home base for the efforts to get the rest of the city up and running again. Engineers worked tirelessly to ensure that the system continued to function properly without becoming overloaded, since it was functioning without the safety net of being connected to the Con Edison grid (Overton 2014). John Bradley, NYU's assistant vice president of sustainability, energy, and technical services, noted that: "The entire neighborhood was dark—everything. And then there was us. It really was a little surreal" (Pyper 2013). He observed that the NYU microgrid supplied essential utilities for the duration of the Sandy event, with no interruptions to key facilities (Overton 2014).

Conclusion

New York University's system was named a "Top Plant" by *Power Magazine* for demonstrating efficiency and reliability in a modern cogeneration plant and associated microgrid (Overton 2014). The university received the U.S. Environmental Protection Agency Energy Star Combined Heat and Power Award in 2013. And during Hurricane Sandy, this system allowed emergency personnel to set up a command post and better assist with storm response (Griffin and Merrihue 2013).

New York University's microgrid provides electricity to 26 buildings on campus, heat to 37 buildings, and chilled water for cooling to 26 buildings. NYU projected that its new cogeneration plant would save the university $5 million to $8 million a year in energy-related costs. The new system emits 23 percent lower carbon emissions, while NO_x and particulate emissions were estimated to drop 68 percent when compared to the original system (Overton 2014).

POWER AFRICA INITIATIVE

In sub-Saharan Africa, up to two-thirds of the population does not have access to electricity, and those who do face chronic blackouts and brownouts (The White House 2013). The International Energy Agency estimates that sub-Saharan Africa will need an investment of more than $300 billion to provide electricity access for the population by 2030 (Burger 2015). Microgrids and other off-grid options have recently become viable for many African communities because of interest from investors and a dip in costs for renewables. These smaller electricity networks can generate enough energy for individual villages or homes using solar or wind energy. Institutions including the African Development Bank are supportive of new renewable-based systems because these systems are cheaper to build than coal-powered plants and the renewable technology aligns with Africa's goals to combat global warming.

The United States has joined the efforts to improve energy accessibility through the Power Africa project. This initiative was led by the U.S. government and committed $7 billion in funds between 2013 and 2018 from 12 government agencies, such as the U.S. Agency for International Development (USAID), and from the World Bank Group, the African Development Bank (AfDB), the Government of Sweden, the Government of Norway, the UK Department for International Development, the International Renewable Energy Agency, African governments, and private-sector partners (USAID 2016, 7). The goal of the initiative is to double electricity access in sub-Saharan Africa within a 5-year period, beginning in 2013. Power Africa focuses on six countries—Ethiopia, Ghana, Kenya, Liberia, Tanzania, and Nigeria—with a goal of adding 10,000 megawatts of power and 20 million new connections (Nixon 2015).

Microgrids in Action

One initiative within Power Africa, the Off-Grid Energy Challenge, is run by the U.S. African Development Foundation (USADF) and offers grants of up to $100,000 to "African companies and organizations providing off-grid solutions that deploy renewable resources and power local economic activities. Challenge winners will have near-term solutions to power the needs of productive and commercial activities, including agriculture production and processing, off-farm businesses, and commercial enterprises" (USADF 2016). USADF, with funding from GE Africa and USAID, reported that as of January 2016, it had awarded 50 entrepreneurs with awards for energy innovation through this challenge. The challenge started in Nigeria and Kenya and has since expanded to nine other countries.

The Off-Grid Energy Challenge aims to bring off-grid energy solutions to unserved and underserved communities through scalable business models. According to U.S. African Development Foundation President and CEO Shari Berenbach, "If you want to reach the majority of Africans, one needs to be thinking about off-grid approaches" (Rucker 2014). Many of the projects that have been funded through this challenge focus on microgrids as solutions. For example, Ethio Resource Group won a grant in 2014 in Ethiopia for a project to install six 1-kilowatt wind turbine systems to power six microgrids. These will provide electricity to 300 households, 20 enterprises, and a health clinic in a rural village in the Amhara region of Ethiopia (USAID 2014). In 2014 in Nigeria, the

group Sky Resources won a grant to develop a small solar microgrid in
Anambra State, where users pay on a per-use basis, supplying electricity
for up to 75 small businesses. In 2016 in Rwanda, RENERG Limited won
a grant to install a 30-kilowatt solar microgrid to generate electricity for
120 households in Muganza Community in Banda Cell, a small village
20 miles from the national grid.

GVE Project Ltd, another winning project, provides two 9-kilowatt
off-grid solar electricity systems, each with two sets of portable battery
kits. Each system supplies electricity to roughly 80 homes and several
businesses, covering more than 2,400 people (Rucker 2014). These systems replaced candles and gasoline, reducing noise and air pollution and
lowering energy costs. Two projects in Tanzania, the SESCOM Kibindu
Biomass Microgrid Project and the Watumia Umeme Cooperative Society Masimbwe Microgrid Project, installed microgrids in local villages:
households in these areas traditionally spent 30 percent of their monthly
income on lighting, and these microgrids contribute to significant cost
savings for the communities. All of these projects aim to support the
local community by providing much-needed electricity access, assisting
schools and small businesses, and reducing the cost of electricity overall.

According to Tom Coogan, regional program director responsible for
Power Africa: "We want to do a number of things: fund innovative solutions, demonstrate successful business models, and showcase technologies that can be scalable, expanded and replicated" (Burger 2015). In the
first 4 years of the Power Africa initiative, Kenya improved its electricity
access rate from 26 to 46 percent, delivering an additional 1.8 million
connections, mostly through urban grid expansion (USAID 2016, 63).

Venture capital companies have substantially contributed to the funding
of microgrid development in Africa. U.S. venture capital raised approximately $38 million in funding for solar companies focused on distributed
solar products in Africa and other emerging markets in the first quarter
of 2015. A Kenyan-based solar company, M-KOPA, sells solar-powered
systems to off-grid houses in Kenya, Tanzania, and Uganda using a mobile
payment system. M-KOPA was able to raise $12.45 million in funding
from various investors, including LGT Venture Philanthropy, an impact
investing firm that aims to improve quality of life for disadvantaged communities (Burger 2015).

Conclusion

Renewable energy and microgrid technology both have a lot of practical potential for powering rural communities in an affordable way. The hope is that the partnerships and off-grid projects that develop through Power Africa will motivate other energy entrepreneurs to continue to expand this new chapter in Africa's energy infrastructure. Experience with small-scale development of rural microgrids may well facilitate the development of microgrids in Africa's fast growing urban areas.

HIGASHIMATSUSHIMA, A MICROGRID COMMUNITY

On March 11, 2011, a massive earthquake occurred off the eastern coast of Japan, causing powerful tsunami waves to hit the country's coast. The earthquake had a magnitude of 9.0, and the resulting tsunami waves affected much of eastern Japan, specifically the Miyagi prefecture. The disaster wiped out entire towns and left millions of Japanese residents without water or power. It caused a meltdown of three reactors at the Fukushima Daiichi nuclear power plant, leading to the shutdown of every nuclear power plant by May 2012 (CNN Wire Staff 2011a). The quake set off a series of aftershocks for the next month, with even a magnitude 7.4 earthquake hitting the Miyagi prefecture nearly a month after the initial quake. About four million homes lost power, and water and rail services were disrupted in some places (CNN Wire Staff 2011b).

At the time of the earthquake, Japan had a total energy generation capacity of 286 gigawatts, 65 percent from fossil fuels and 17 percent from nuclear power (Marnay et al. 2015, 51). The nation's electricity was transmitted on two large and one smaller grid. After this disaster, Japan's energy supply system was called into question, and development began for a system that could withstand extreme weather events. The Japanese government, utility companies, and researchers began to analyze damage and assess where infrastructure could be improved. According to a survey of 571 businesses by the Japanese government, a majority of businesses view blackouts to be the most influential factor threatening business continuity (Marnay et al. 2015, 55).

Microgrids in Action

The Japanese government has long seen microgrids as a vehicle for achieving high renewable penetration in electricity supply. Japan had several microgrids in operation during the time of the earthquake and tsunami, and two of the microgrids even functioned as secure islands after the disaster, demonstrating that localized energy systems with well-developed lifelines could be critical for emergency service continuity (Marnay et al. 2015, 57).

The Japanese government established the "Future City" Initiative in 2012 to develop solutions for disaster preparedness and environmental improvement. The cabinet chose Higashimatsushima to be a model city for this initiative, and an organization called the Higashimatsushima Organization for Progress and Economy, Education, Energy (HOPE) was established through private and public collaboration to lead the effort (JICA 2015).

The city of Higashimatsushima suffered the worst flood damage of all cities in the Miyagi prefecture in the aftermath of the earthquakes and tsunami, leaving the city 65 percent underwater, killing 1,100 people, and displacing 10,000 more citizens from their homes. To establish a more sustainable and resilient energy infrastructure as a part of its recovery and resilience plan, the city of Higashimatsushima set a goal to be a "Net Zero Energy City" by 2022, so that the entire city runs on locally produced energy (Movellan 2015; Willmott 2015).

Importantly, the city is building Japan's first microgrid community, the Higashimatsushima Disaster-Prepared, Smart Eco-Town, in collaboration with Japan's leading housing development company, Sekisui House, with research and funding from the Ministry of the Environment. The community consists of 70 single-family homes and 15 multifamily apartments, which are owned by the city and rented to families who lost homes because of the tsunami. The homes will be steel-framed, earthquake-resistant, and heavily insulated for maximum energy efficiency (Movellan 2015). In the event of a disaster, the community microgrid will also be used to supply power to hospitals and other households (Kaneko 2015). According to solar energy expert Junko Movellan (2015):

Energy loads at this community will be served by the integration of distributed, clean energy—PV systems (470 kW) and bio-diesel generator (500 kW) together with a large-scale energy storage (500 kWh). When the centralized power grid becomes unavailable, the community will be able to function autonomously. There are three PV systems in this community: a 400-kW PV system over a reservoir, a 60-kW system on the apartment buildings, and a 10-kW system at the assembly hall, which serves as a community-gathering place in case of emergencies. Any excess electricity generated from these PV systems during the day will be stored in the battery system and used at night. When disconnected from the traditional grid, the town can supply three day's worth of everyday energy needs for residents and buildings in the town. During a prolonged power failure, the town can still provide minimum energy needs for the hospitals and assembly hall.

A unique element of the city's electrical grid infrastructure is that it is owned and operated by the city, which has a contract with the regional utility company (Movellan 2015). The system will use a community energy management system (CEMS) to bill customers, which allows for monitoring of generation and consumption, assesses peak-demand times, and helps use energy more efficiently within the community (Haugner 2015).

Higashimatsushima decided to pursue sustainable energy such as wind, solar, and biodiesel energy generation coupled with large-scale energy storage. In addition to microgrids, the city has pursued a number of projects to reduce dependence on the grid. For example, one initial project was using a former park, which was greatly affected by the floods, to create a 2-megawatt solar photovoltaic project, which was completed in 2013 by the private company Mitsui & Co., Ltd. (Movellan 2015). The company also began work on building a mega solar power facility. Mitsui's total investment in support of the solar project was approximately 1.24 billion yen (about US$12 million) (Mitsui & Co. 2012).

The city continues to strategize and develop goals for recovery and resilience to both bolster the local economy and combat climate change. In order to reach the goal of providing 100 percent of the electricity for the city by 2022, Higashimatsushima must install 33 to 44 megawatts of renewable energy capacity. The city continues to explore options for renewable energy sources, such as biomass generation, or fuel made from organic waste (Movellan 2015).

Conclusion

The city of Higashimatsushima is making progress in its effort to transform its energy system. While the origin of the effort can be found in nuclear disaster, forward movement has been made possible by collaboration between the city-owned electrical grid and private companies, facilitated by the Ministry of the Environment. Multiple projects have been completed and are operational. The microgrid community has the potential to be a model system for other cities in Japan.

ANALYSIS OF SUSTAINABLE ENERGY: MICROGRID DEVELOPMENT AT NEW YORK UNIVERSITY, THE POWER AFRICA INITIATIVE, AND HIGASHIMATSUSHIMA IN JAPAN

Sustainable Energy as an Issue of Values

Almost every part of our lifestyle requires energy: lights, heat, air-conditioning, transport, entertainment, food, water, sewage, communications—the list is nearly endless. We aren't born requiring an iPhone. Over time, we learn a system of values and beliefs that influences our view of how the world works and our preferences for goods, services, relationships, and other choices; in this case, resulting in our use of energy. The need for energy does not imply the use of fossil fuels or renewables; another set of values will influence our preferences (if we have any) for the type of energy we use. It also does not imply a concern for energy efficiency; another set of values that may or may not be connected to the preference for renewable energy influences the preference for energy efficiency. The value construct that leads to microgrid adoption may be more related to the desire to reduce costs, prevent power loss, and modernize a critical piece of infrastructure than to any concern about climate change or environmental degradation.

But the key value dimension related to energy is its centrality to our lifestyle. Our need for energy affects many different aspects of our life ranging from local politics, to consumer choice, to foreign policy. Threats to energy supplies are threats to our way of life. To the extent that microgrids are presented as methods for enhancing the security and efficiency of energy, we can expect to see deep support and a willingness to pay for such enhancements.

Sustainable Energy as a Political Issue

The centrality of energy was clearly demonstrated more than four decades ago when OPEC embargoed oil bound for the United States resulting in higher prices and long lines at gas stations. Cries for "energy independence" began then and continue today. Energy politics is also a central dimension of the politics of climate change and environmental protection. Burning fossil fuels emits greenhouse gases, and extracting fossil fuels from the earth can damage marine and terrestrial ecosystems. Finally, public investment in energy infrastructure such as smart grids raises the issue of the role of government in the economy and related issues of taxation and user fees.

Sustainable Energy as an Issue of Science and Technology

The energy issue is essentially an issue of science and technology. The move from human energy to other forms of energy has been the result of scientific advances. Energy efficiency and distributed generation of energy is facilitated by computer-controlled smart grids. Improvements in renewable energy technology and in energy storage will make the transition to a sustainable economy feasible. The environmental problems caused by energy technology will eventually be addressed by new energy technology.

Sustainable Energy as an Issue of Public Policy Design

If we are to modernize the electrical grid, it will be financed by either government taxpayers or electric utility rate payers—essentially the same people. Let us assume that the technical capacity to enable distributed generation of energy and net metering is established, and that household electric meters can go forward or backward and energy can be used or stored to maximize efficiency. Without effective regulation of utilities and the price of electricity, the system may still function poorly. Reducing dependence on fossil fuels will require smart grids but will also require that a utility's investment in outmoded fossil fuel plants is recovered and its debt retired. Without that policy design, utilities will continue to burn fossil fuels in inefficient plants or go bankrupt.

The rate structure, which is a key element of energy policy design, can also be used to encourage behaviors from consumers of energy. Efficiency and off-peak utilization of energy can be influenced by the rate structure. Much can be accomplished with creative policy designs influencing rate structures, utility profit, and the availability of capital for energy infrastructure.

Sustainable Energy as a Management Issue

Transitioning to a new energy system will take time, money, technology, political will, creative policy, and rate structures and organizational capacity. New technologies require newly developed and trained technical experts. The transition requires coordination, operational information, motivation, and leadership. In the final analysis, a new energy system must be built and then maintained. This requires management skill and deep understanding of the process of organizing change.

Sustainable Energy as a Multidimensional Issue

There are many factors that make energy a multidimensional issue, but my view is that the dominant themes are technology, the policy design of rate structure, and incentives for private investment. The transition to renewable energy must result in lower cost and more reliability if it is to disrupt the existing system. Energy is too important to play around with. Yes, it might be better to have a sustainable renewable energy system—but not if it is an expensive and unreliable system. If we can bring down the price and make the system more efficient and cost-effective, renewable energy will be a by-product, but not the causal agent. More efficient solar cells, better battery storage, and transmission systems that are decentralized and have backup capabilities are all part of a lower cost, more reliable energy system.

9

PARKS AND PUBLIC SPACE

Parks are a valuable resource in the sustainable city. In densely populated areas, where many people live in small spaces without outdoor private space, cities must create shared open space that everyone can use. And when land is scarce, they must figure out innovative and cost-effective ways to do this. This case looks at four parks/open space plazas in four different cities—three in the United States and one in Brazil. What these parks have in common is that they each were transformed and repurposed from a nonfunctional, abandoned site to a widely popular open space that everyone can utilize.

GAS WORKS PARK, SEATTLE, WASHINGTON

Gas Works Park is a 20.5-acre park located north of downtown Seattle, Washington. The park, extending 400 feet into the north shore of Lake Union with 1,900 feet of shoreline, offers unique views of the Seattle skyline. The landscape of the park consists largely of rolling hills along with an old gas plant. The Gas Works Park site was once a garbage-burning site and most recently was a gas manufacturing plant that supplied the city for 50 years by converting coal and oil to gas. The Gas Works plant was closed in 1956 because the technology of manufacturing gas was outdated and more expensive than alternatives. The abandoned plant was considered an eyesore in the neighborhood, and by 1963 the city was working to purchase the site. What eventually became the park was open to the public in stages between 1973 and 1976. The park is owned and operated by the City of

Seattle Department of Parks and Recreation. Friends of Gas Works Park is a nonprofit community organization that was formed in the early 2000s and is responsible for stewardship of the park.

The city's Department of Parks and Recreation bought the land from Gas Light Company for $1.3 million in order to turn it into a traditional landscaped park. The deal stipulated that the gas company would clear the site by the transfer date. The city allocated $1.75 million as a Forward Thrust Bond Issue to transform this highly polluted site into a thriving urban space. In 1970, architect Richard Haag was commissioned to plan the park. Haag's plan was to keep the industrial plant on-site to incorporate the history of the site into the new park. But this wasn't well received by city council members or the public, who wanted the industrial plant removed. After the poor reception of his vision for the site, Haag began a campaign to advocate for his vision of the park (Bennett and Teague 1999).

His campaign focused on the "geographic location, historic significance and esthetic resources" of the site. Haag envisioned the park as an urban area where users could extend their city experience instead of merely escaping from it. He saw the old "Victorian" park plans as anachronistic and limited in vision. Haag worked to persuade the public that his concept still met the goal of a beautiful public park, while recognizing the postindustrial ruins. He offered to meet with community members at the site to describe his vision in person and persuade the community to incorporate the structure that had once allowed Seattle to prosper and expand. Haag's vision preserved an industrial feel and allowed natural systems to coexist with urban ones (Way 2013, 28). The Seattle City Council approved Haag's master plan in 1971.

Haag began experimenting with the then-emerging field of bioremediation, the process of using microorganisms or other forms of life to consume and break down environmental pollutants, in order to clean up a polluted site. The ground of the Gas Works site was saturated with tar and aromatic hydrocarbons. Tests estimated industrial deposits coated the ground as deep as 9 feet in some areas. Land that had once thrived with marshes and grass was now barren because of contamination (Way 2013, 31). To begin, the soil was tilled to disrupt the anaerobic environment of the soil. Then sewage sludge, sawdust, biomass, and leaf litter were added to begin a passive cleansing process. Within the first year of work, grass began to grow. The most contaminated soil, along with construction rubble that could not be recycled, was put aside to form the base of Kite Hill, which was capped with 18 inches of hard-packed clay. The rolling topography of the park encouraged the dynamic breakdown of pollutants throughout the site. The hills ensured that stormwater was always flowing, while in the valleys the water rested and seeped into the ground and replenished the cleansing process. Haag's landscape design created a self-sustaining environment that greened itself allowing for the land to be once again productive (Way 2013). The old towers of the plant are now iconic features of the park. The old boiler house was converted into a popular picnic area, and the former exhauster-compressor building was transformed into an open-air play barn (The Cultural Landscape Foundation 2016).

However, the park has experienced challenges with continued contamination. In 1997, tar began to bubble up on Kite Hill, and city officials had to haul away around 3,000 pounds of tar from the park. Despite the work of city officials and the natural, passive cleaning process under way, in 2001 two major problem areas were identified. Benzene, a carcinogenic by-product of the gas manufacturing process, was found in the groundwater in the southeast corner of the park. A pipe system was installed to remove it from the groundwater. Naphthalene, another by-product, was found in the form of underground plumes in the southeast corner of the park. These deposits have been continually sinking underground for years to sit under Lake Union, and ecologists continue to monitor this process. The 2001 cleanup project cost an estimated $3 million (Clarridge 2001).

Similar contamination issues were uncovered and required remediation in 2005, 2012, and 2014. Site cleanup required adding about 2 feet of topsoil to the area as well as geotextile fabric. The 2014 project focused on improving the groundcover of Kite Hill to preserve the structure's shape; Kite Hill closed for nearly a year. An in-water cleanup of contaminated sediment is planned. Puget Sound Energy (the new owners of Washington Natural Gas) and the City of Seattle are responsible for paying for these cleanup efforts (State of Washington 2014). The cleanup of toxic waste represents an important challenge when restoring an industrial site.

Richard Haag's design for Gas Works Park reimagined a toxic industrial space into a healthy public space. It was listed as a Seattle Historical Landmark in 1999 and is listed on the Washington State Historic Register and the National Register of Historic Places. The park was considered radical because of the way that it transformed contaminated land into land fit for public use. The park was among the first postindustrial spaces to be transformed for public use without complete removal of its pollutants (Way 2013, 30). Gas Works Park was named a national landmark in 2013, and it has become one of the most popular parks in Seattle (The Cultural Landscape Foundation 2016).

HIGH LINE PARK, NEW YORK CITY, NEW YORK

High Line Park in New York City is an elevated (30 feet high) former freight rail line located on Manhattan's West Side that was transformed into a 1.5-mile-long public park. The park is owned by the City of New York and is maintained, operated, and programmed by the nonprofit conservancy Friends of the High Line in partnership with the city's Department of Parks and Recreation. The building of the original High Line elevated roadbed was part of the West Side Improvement Project and was motivated by public discontent over citizens being injured and killed by the freight trains that traveled up and down 10th and 11th Avenues (Berg 2015). The original High Line operated between 1934 and 1980 and was designed to carry goods by rail directly from the West Side docks to nearby factories in Manhattan. Its trains carried meat, raw goods, and manufactured products to and from the city's factories. When containerized shipping was introduced, the factories and port died off and the High Line sat unused and largely abandoned.

Friends of the High Line was founded in 1999 as a 501(c)(3) nonprofit by Joshua David and Robert Hammond, residents of the neighborhood, to advocate for the High Line's preservation in the face of demolition proposals. The pair hoped to emulate the Central Park Conservancy model, utilizing a public-private partnership to maintain and operate a public park. The city originally wanted the High Line demolished in order to spur economic and real estate development in the area. Local real estate interests, such as Chelsea Property Owners (CPO), also wanted the line demolished. The prospects for the park project improved when Mayor Bloomberg took office in 2002, since he supported this project as an economic development opportunity (Bowen and Stepan 2014).

Planning for the High Line's preservation began in the early 2000s, and there was an open competition for its design. In 2005, the city rezoned the area to create the Special West Chelsea District, to allow a mix of residential, commercial, and manufacturing industries (Bowen and Stepan 2014). The rezoning included a provision that any developer could build higher than the zoning allowed, for a $50 per square foot fee, well below market rate. It also created transferable development rights (TDRs) areas, allowing private property owners to sell their new, unbuilt TDRs (or "air rights") to noncontiguous landowners. These measures would create tax revenue to offset the city's renovation costs for what came to be High Line Park. CSX Transportation, owner of the actual railroad structure, donated the High Line to the city in exchange for release of all liabilities. Community members were concerned that the park project would lead to gentrification but were satisfied when Friends of the High Line made changes to the proposed zoning, setting aside 30 percent for affordable housing.

Construction began in 2006, and the first section of the High Line opened to the public in June 2009. Just before the opening, New York City and Friends of the High Line signed a formal agreement, ensuring that the park would become part of the city's public park system, but Friends of the High Line would operate and manage it and provide at least 90 percent of the operating budget (Bowen and Stepan 2014). Friends of the High Line now provides 98 percent of the High Line's annual budget, which averages $3 million a year for operations and maintenance, plus about $2 million to $3 million for overhead and staff. The park is more expensive to run and maintain than many other parks in the city. The capital construction of the first two sections of the park cost $152.3 million; $112.2 million came from the city, $20 million from the federal government, and $400,000 from New York State.

By 2013, the city's analysis put the economic benefit of the park at close to a billion dollars. Accounting for the city's investment of about $115 million, the city's net benefit over 20 years will be more than $900 million, or a return of more than 800 percent (Broder 2012, 250). According to a

Friends of the High Line blog post: "Today, as you stroll through the park, you will see remnants of the High Line's past function as an active railway preserved in the park's design. Spurs where the trains once pulled into neighboring buildings to load and unload their freight are now horticultural preserves. The High Line's original rail tracks are embedded within the planting beds. You can even spot old meat hooks hanging along the factory buildings near the southern end of the High Line" (Harvey 2015).

The park design and operation also incorporates elements of sustainability. The planting design was inspired by the self-seeded landscape that naturally grew on the High Line's roadbed before it became a park. The perennials, grasses, shrubs, and trees were chosen for their hardiness and sustainability, with a focus on native species. The plant selection favors native, drought-tolerant, and low-maintenance species, reflecting the original microclimates of the High Line, and cutting down on the resources that go into the landscape (Friends of the High Line 2016). The park itself functions as a green roof. Porous pathways contain joints so water can drain between planks, reducing the amount of stormwater that runs off the site into the sewer system. The drip irrigation system is designed to allow planting beds to retain as much water as possible. There are also on-site composting facilities that help process garden waste into compost.

High Line Park was made possible by sophisticated land-use planning by the city, mayoral support, innovative political and public relations strategies, and significant financial contributions by private interests (Broder 2012, 246). By July 2014, the park had hosted more than 20 million visitors and today features 450 programs and activities each year (Friends of the High Line 2016). It is widely viewed as a successful example of a public-private partnership, utilizing unused space and stimulating a real estate and development boom in the neighborhood.

VICTOR CIVITA PLAZA, SÃO PAULO, BRAZIL

Victor Civita Plaza is located in eastern São Paulo in the wealthy Pinheiros district. The plaza is built on top of 130,000 square feet of an old, highly contaminated medical waste incinerator (Dietzsch 2015). The idea to rehabilitate this space came from the São Paulo–based publishing company Editora Abril, founded by Victor Civita, and whose headquarters are located near the site (Pompeu de Toledo 2016). The company formed a partnership with the city to reclaim this space and make it available for public use. The city was in the midst of its "100 Municipal Parks" plan to add new parks to the city (Martignoni 2009). While the original plan for the plaza was to cover the contaminated soil with fresh soil and use a traditional design, the architects of the space decided that this would diminish its teaching potential. However, their plan was unprecedented; there were no direct laws that dealt with the repurposing of contaminated areas, and the designers had to generate political support to sell the project (Pedersen 2009). Adriana Levisky, a local architect who specializes in zoning and planning, is credited with the political savvy to coordinate stakeholders, which made this park a success (Archdeacon 2000). After extensive planning and debate, the designers were allowed to incorporate the site's polluted history into the architectural design, and the park was completed in 2008 (Dietzsch 2015).

Instead of completely covering the contaminated soil with fresh soil, a 3-foot-tall steel framework was constructed to support a floating deck made out of recycled Brazilian hardwood, which sits above the contaminated soil. The 3-foot-high cap was required by the city's Sanitation Department in order to ensure the contamination was contained before dedicating the site to public use. The deck is built around preexisting elements of the space, and it creates a walkway for visitors to stroll through the park while they interact with the old and new elements of the space. On an uncovered area of the plaza, the landscape designers constructed a Tech Garden on a series of elevated trays; the Tech Garden is meant to serve as a natural filtration system by storing and cleansing rainwater (Pedersen 2009). Coconut fiber–lined pipes allow plants to draw water when needed, making the garden self-regulating. Furthermore, plants were selected for their educational potential: the plants in the garden demonstrate processes like soil purification, biodiesel generation, and hydroponics (Pedersen 2009). Along with efficient water use, the plaza also rations energy use with solar panels, LED lighting, and a design that promotes the use of natural light (UN Global Compact 2016; Yavar 2014).

The design of Victor Civita Plaza is also purposely socially inclusive. The park's goal was to make an open and free space that would foster environmental awareness, sense of community, and cultural growth (Yavar 2014). According to one of the park's designers, "the idea was to create the whole site as if it was an open museum" (Dietzsch 2015). The deck includes a shaded stage area where various types of events are held, such as concerts, theater, lectures, and classes. The gardens contain educational panels that tell about the species of plants and their uses. Inside the old incinerator building is a sustainability museum that explains the history of the site, and there is further space for educational opportunities (UN Global Compact 2016). These spaces feature art exhibits from the São Paulo Art Museum, as well as children's environmental educational workshops. A housing unit for elderly people is located adjacent to the plaza and is fully integrated into the activities presented there (UN Global Compact 2016).

Victor Civita Plaza is a public space that pushes the boundaries between architecture and the environment. It is also an example of public-private partnership between the city and Editora Abril. In a developing city that has little open public space, Victor Civita Plaza demonstrates how contaminated industrial sites can fit into the sustainable city (Pedersen 2009). This public space is alive and interactive and allows visitors to understand the many layers of its previous use. As one of the designers of the plaza stated, Victor Civita Plaza demonstrates "the importance of relatively small interventions that could elucidate the complexity of the city and could bring people closer to the understanding of more harmonious and sustainable design solutions" (Dietzsch 2015).

CANAL PARK, WASHINGTON, D.C.

Canal Park opened to the public in late 2012 in what used to be a parking lot for school buses in the rapidly growing Capitol Riverfront neighborhood of Washington, D.C. (Neibauer 2013). It is one of many spaces being redesigned in the business improvement district in an effort to create a "smartly designed" neighborhood full of "inspired workspaces, open parks, walkable streets, and emphasis on nature" to "promote innovative workstyles and healthier lifestyles" (Capitol Riverfront BID 2016a). The three blocks that make up Canal Park were once part of the historic Washington City Canal that connected the Anacostia and Potomac Rivers. Canal Street was created in the early 1900s when the canal was paved over. Over the past century, Canal Street was owned by the District government and used for various purposes. By the 1990s, plans were proposed to build a park on Canal Street, but it never materialized; the neighborhood at this time had few of the amenities it has now and was much less developed (Wheeler 2012). It wasn't until 2000 that real effort to reimagine this space occurred; at this time, local developer WC Smith created the Canal Park Development Association (CPDA), a nonprofit organization, to secure the site and oversee development. The CPDA partnered with the Anacostia Waterfront Corporation and the District's Deputy Mayor for Planning and Economic Development to start plans for a park in that site. Various features of the park were a result of a public design competition sponsored by this partnership. Construction for the park finally began in 2010 (Capitol Riverfront BID 2016b).

Canal Park is a $26.5 million public-private project—it is publicly owned but privately managed. The federal government awarded a $2.5 million grant to the Canal Park developers, and this grant was matched by one of the developers involved in the project. The District government funded the bulk of the project, providing $13.5 million. The D.C. Housing Authority was able to provide $3.9 million in funding as part of a new market tax credit to bring housing to the area (Wheeler 2012). The Canal Park Development Association and the Capitol Riverfront Business Improvement District both manage the park and all activities and events held there. The park is centrally located to add economic and social benefit to an area with governmental, residential, and commercial interests. It is an accessible public space with three metro lines nearby and an example of how innovative and inclusive public spaces can serve as catalysts for development (Capitol Riverfront BID 2012). Along with the building of the park on Canal Street, the city also commissioned 1,800 mixed-housing units to replace 700 public housing units (Wheeler 2012). The Capitol Riverfront neighborhood is home to many jobs, including the main office of the District's Department of Transportation, and to more than

4,000 residents, and the neighborhood receives more than 2.5 million visitors yearly (Capitol Riverfront BID 2012).

The park's design reflects the area's history, including a linear rain garden and three pavilions meant to look like the floating barges that once drifted along the canal. The park's design also embodies the three pillars of sustainability; it was designed to be a social gathering place, an economic trigger, and an environmentally conscious site. The pavilions, one of which houses a 65-person restaurant, are made from recycled materials and help to minimize the park's energy footprint by including geothermal heating and cooling mechanisms as well as green roofs and an electric vehicle charging station (Capitol Riverfront BID 2016).

An impressive aspect of this site's sustainable landscape design is its comprehensive stormwater management and water reuse system. The linear rain gardens help to collect and naturally filter stormwater that eventually drains into the park's extensive underground system. Runoff is collected in cisterns that can hold up to 8,000 gallons of water collected from the park and the bordering blocks. This system meets 95 percent of the park's water needs, including water for irrigation, the ice-skating rink, and the fountains (Capitol Riverfront BID 2016). This system also contains the geothermal wells that reduce the park's energy consumption by almost 40 percent (Lewis 2013). Even though the site was once a toxic brownfield, the native plant habitat was reintroduced to the area when the park was being built, and low-impact-design tree pits and bio swales were used to increase the park's water filtration capacity. Canal Park has received both LEED and SITES Gold certifications, the former for building design and the latter for sustainable landscapes (Sustainable SITES Initiative 2016; USGBC 2009).

The park's most popular feature is undeniably the outdoor, 10,000-square-foot linear ice-skating rink. In the summer months, the rink is converted into a 42-jet interactive fountain, also very popular among residents and visitors (Wheeler 2012). The park contains seating spaces along the walkways, green spaces, and the pavilions, as well as interactive sculptures designed by local artist David Hess. The Capitol Riverfront Business Improvement District keeps the park booked with events, including various outdoor gym activities, an outdoor movie series, music festival, and farmers' markets.

ANALYSIS OF PUBLIC SPACE: THE HIGH LINE IN NEW YORK CITY, VICTOR CIVITA PLAZA IN SÃO PAULO, CANAL PARK IN THE DISTRICT OF COLUMBIA, AND GAS WORKS PARK IN SEATTLE

Public Space as an Issue of Values

When someone by choice or compulsion settles in a city, that individual makes the decision to limit the amount of private space he or she either owns or has access to. Personal space in cities is measured by the square foot; in the countryside it is measured by the acre. But the human need for open space, vistas, recreational space, and a psychological sense of space remains. But how much space and in what form is a matter of individual and community values. Increasingly, we see excellent public space as a key piece of urban infrastructure, necessary for urban sustainability. Interesting and beautiful public spaces raise the price of adjoining land and structures, provide destinations for tourism and recreation, and provide a place for social interaction and public gatherings. These are all key urban activities related to the values that bring people into cities. Public space facilitates and democratizes that interaction.

Public Space as a Political Issue

All public resources, facilities, and infrastructure are inherently political. Who pays for them, how they are paid for, where they are located, and who benefits are all potentially issues of political contention. The type of public space can also be a subject of intense debate: a nature preserve versus a golf course, a basketball court versus a public garden. It's not hard to envision the potential conflict. Then there is the overall issue of public spending on parks versus spending on other needs such as schools, roads, and police. National parks in the United States are not urban public spaces, but they are frequently subject to intense opposition by local interests even as their creation is supported by a wide national consensus. Land is a scarce resource, and as the number of people on the planet grows and the amount of land remains the same, we can only expect the conflict to become more intense.

Public Space as an Issue of Science and Technology

Unlike many other areas of urban sustainability, public space is not really thought of as an issue of science and technology. There are exceptions as spaces that were once considered inaccessible because of toxics or other land uses can be remediated and used for parks. Technology can be used to maintain and renovate parks, but for the most part, science and technology is not a central determinant of the quality and quantity of urban parkland.

Public Space as an Issue of Public Policy Design

One of the more interesting elements of public policy related to space in the past several decades has been the use of public-private partnerships to develop and manage public space. In exchange for concessions to private developers—to develop more square footage than normally permitted or in exchange for other public benefits—the private developer might invest in the development and/or maintenance of public space. This can enhance the value of the developer's private property by creating a public amenity and by increasing the financial return from a development. It can help enhance a firm's image with the public and provide a more attractive project. From the public's perspective, it can provide a low cost or even free public space. This is of particular value if the private sector assumes the costs of maintaining the park. While there are examples of private spaces that have not been well designed or where their public purpose was not clear, there are many examples of successful private plazas and spaces that are indistinguishable from public spaces. Many college campuses and nonprofit museums have these types of spaces, and they can be beautiful and lively places.

There are examples of public spaces taken over by private nonprofit organizations. The most famous and successful is Central Park, which was brought back from the dead and has been successfully run by the nonprofit Central Park Conservancy for decades. High Line Park in New York City was developed and managed by the nonprofit Friends of the High Line. It too has been a spectacular success. Throughout the United

States there are a wide number of groups that raise funds for parks and help support them, even if they do not hold a contract and manage them. The tax code of course encourages these types of organizations.

The mix of public and private space is always an issue of policy design, and even the concept of private space is a policy construct. In legal terms, a sovereign state always retains rights over private property, which can be taken by government for public purposes through government's power of eminent domain. In a more urbanized world, the importance of public space will continue to grow as will its importance as an issue of public policy.

Public Space as a Management Issue

Public space requires capital to plan and develop, and then it requires resources and organizational capacity to maintain and improve. Space is not simply set aside: if it is used it requires landscaping, waste removal, and repairs. It is also more politically attractive to cut the ribbon on a new park than to mow the lawn in an old park. If a public space is not put to use and does not appear popular and dynamic, it can enter a descending spiral of neglect, disrepair, and even destruction. A public space must be seen as more than a place: it must be seen as a facility.

Managing many public spaces as an integrated system is a complex task requiring planning, market analysis, and a high level of project management capacity. It requires political skill to navigate because by definition, some neighborhoods will end up with better park access than others. It also must be integrated with other urban infrastructure such as transportation. A popular park on a Sunday afternoon will not stay popular for long if subway access is closed for the weekend. Public safety is another key service area that requires the capacity for a high degree of interorganizational coordination.

Public Space as a Multidimensional Issue

With the exception of science and technology, public space is an issue of every other dimension of our analytic framework. Siting parks raises political issues; the use of parks generates issues of values. The role of the private sector and the commercialization of the park and surrounding areas is also a key issue. A fundamental factor of space is that even though we are making more people, we are not making more of our planet. As time goes on, the issue can only become more contentious. As we become more urban, public space becomes an ever more essential public service.

10

SUSTAINABLE URBAN LIVING

Sustainable urban living is about more than energy-efficient buildings, mass transit, and green spaces. A growing aspect of sustainable urban living is the "sharing economy." Sharing has always been a part of urban life (e.g., public libraries and community spaces), but in the past few years, cities have seen a significant revival and acceleration in sharing activity and innovation. In cities around the world, people are now welcoming guests into spare rooms, sharing tools and equipment, and paying for rides in cars of people they don't know. According to a National League of Cities (NLC) survey, more than half of cities in the United States indicated that they have seen some growth in the sharing economy, with 16 percent classifying this growth as rapid (DuPuis and Rainwater 2015, 5). The size and scale of leading companies operating within the sharing economy, most of which didn't exist a decade ago, now rival that of some of the world's largest businesses in transportation, hospitality, and other sectors.

From innovative technologies and business models to redefined concepts of equity and safety, the sharing economy is starting to change our cities. Cities have many resources that can easily and effectively be redistributed and shared. By allowing people to consume less and own less, thereby using fewer resources, the sharing economy promotes urban sustainability (Hirshon et al. 2015). The sharing economy can lead to more sustainable consumption while remaining compatible with economic growth. According to Kurt Matzler and his colleagues, the sharing economy "seems to hold the potential to unite cost reduction, benefit augmentation, convenience and environmental consciousness in one mode of consumption" (Matzler, Veider, and Kathan 2014). However, sharing economy services have also presented cities with unprecedented, complex questions. The greatest challenge for cities is finding a balance between embracing these platforms, as well as the various benefits they offer to residents and visitors, and regulating them to ensure safety and responsibility. With more types of sharing businesses entering the market, and the rising popularity of these new applications and services, city leaders have been forced to address these issues all at once (DuPuis and Rainwater 2015, 5).

In this case study, I will look at two mainstream platforms within the sharing economy: Airbnb and Uber. Both companies have achieved success by democratizing and creating access to two of the largest assets people have: their homes and cars. Airbnb is now valued at $25 billion, rivaling large hotel chains, while Uber is valued at $65 billion, similar to major car companies (Slee 2016). I'll provide a short overview of the sharing economy, then brief backgrounds of both companies, including the benefits and challenges presented in cities.

WHAT IS THE SHARING ECONOMY?

The sharing economy, or collaborative consumption, "is the peer-to-peer–based activity of obtaining, giving, or sharing the access to goods and services, coordinated through community–based online services" (Hamari, Sjöklint, and Ukkonen 2016, 2047). It is a system built around the utilization of unused or underused resources (Bond 2015, 77). The sharing economy dates back to the 1990s with the founding of online marketplaces eBay and Craigslist, which allow for the recirculation of goods. But the modern sharing economy looks slightly different, fueled by information and communication technology and the proliferation of Web–based communities, stemming from the tech-driven culture of Silicon Valley (Hamari, Sjöklint, and Ukkonen 2016, 2048). According to an analysis by PricewaterhouseCoopers (PwC), by 2025 the five main sectors of the sharing economy—transportation, accommodation, finance, music/television/video streaming, and online staffing—could represent $335 billion in revenue worldwide (PwC 2015, 14).

The rapid growth of these peer-to-peer platforms has been enabled by innovations in technology, which have streamlined entry into the market for suppliers, facilitated easy access to searchable listings for consumers, and kept overhead costs low (Zervas, Prosperio, Byers 2016). As a result, sharing assets is cheaper and easier than ever before and, therefore, possible on a much larger scale. Before the Internet, renting a good or space from someone else was feasible, but not easy to do. Now websites match up apartment owners and renters; smartphones with GPS let people see where the nearest rentable car is parked; social networks provide a way to check up on people and build trust; and online payment systems handle the billing (*Economist* 2013). Just as YouTube disrupted television and blogs compete with mainstream media, the sharing economy replaces the industrial model of companies owning and people consuming and allows everyone to be both consumer and producer.

The sharing economy has emerged as a way for people to utilize an item or service without having to own it. The idea of ownership is changing—today's generation doesn't see ownership as a privilege, like their parents did, but sometimes as a burden. A 2011 survey completed by BAV Consulting showed that 66 percent of consumers (and 77 percent of millennials) prefer a lifestyle with fewer possessions (PwC 2015, 14). We're moving away from a world organized around ownership to one organized around access to assets. Instead of showing off possessions, people now want to show off the places they have traveled to and the experiences they have had. The sharing economy allows people to spend less and still have quality experiences. In some cases, it even allows them to gain access to new services they wouldn't have been able to afford previously. This new trend of "dis-ownership" gives individuals and families more choice and opens the door to a more efficient lifestyle. One example I've heard frequently is that a power drill is used, on average, for only 12 to 13 minutes in its entire life. So why buy one when you can rent or borrow it instead (Botsman and Rogers 2010)?

A growing concern about climate change and future sustainability has made the sharing economy an appealing alternative for environmentally conscious consumers. With sharing, less energy is needed for transportation and production of goods, and less waste is created as everyday products and services are shared among a group (Belk 2010). Some research has indicated that participation in the sharing economy can be ecologically sustainable (Prothero et al. 2011; Sacks 2011). Sarah Cannon and Lawrence Summers found that car sharers report reducing their vehicle miles traveled by 44 percent while surveys in Europe show CO_2 emissions are being cut by up to 50 percent (Cannon and Summers 2014). However, to date, there are no comprehensive studies of the environmental impact of the sharing economy (Schor 2014).

While some cities are embracing the technology, innovation, and new business models that the sharing economy brings, others are resistant to change or have concerns about the safety and responsibility of these new services that aren't regulated under the same conditions as traditional services. The biggest challenge to cities is regulatory uncertainty. The explosive growth of the sharing economy has led to regulatory and political battles in cities all over the world. Governments do not currently have a strong regulatory framework to govern these new services (Posen 2016; Quattrone et al. 2016). Regulating the sharing economy is challenging because existing laws were developed without considering the new issues it presents. These types of companies do not fit industry regulations perfectly and sometimes operate outside the law. Critics, such as yellow-taxicab drivers, landlords, and hotel owners, "argue that operating without regulation gives start-ups an unfair advantage over highly regulated incumbents" (Chang 2014). The city is also forced to respond quickly, with little time or opportunity to develop an effective response, sometimes enacting legislation without substantial stakeholder participation. Policy makers are often under pressure to permit the services their constituents value while also ensuring safety, fairness, and the best interests of the community. According to a recent report, "Many municipal governments are attempting to impose old regulations on these new marketplaces without much thought about whether these laws apply to these companies, and without a complete understanding of the benefits and drawbacks generated by these new services" (Quattrone et al. 2016). There are also questions about the enforceability of new rules. Many sharing services are relatively invisible. In any case, the sharing economy is here to stay, and rules should be designed to facilitate rather than discourage its development.

RIDE SHARING: UBER

Uber, which I discussed briefly in chapter 3, is a service that offers on-demand transportation access through a cell phone application. It provides an excellent example of poorly thought through efforts by government to regulate the sharing economy. Uber was created by tech entrepreneurs Garrett Camp and Travis Kalanick in May 2010 in San Francisco, California, with $200,000 in seed funding. The company originally started as a luxury black car service and launched a lower-cost service called UberX in 2012 that contributed to the company's rapid and widespread growth. Uber currently operates in more than 500 cities worldwide. Between 2009 and 2015, it received $6.61 billion in seed funding from 52 investors (Jin et al. 2016). At this writing, it was valued at roughly $51 billion (Woo and Bales forthcoming).

Uber does not call itself a transportation service. Rather, it is a technology platform that allows users to easily access a ride through a third-party transportation provider. Uber is "in the business of delivering cars." A customer can use an application on his or her phone, request a ride to an exact location, get a fare estimate and estimated time of arrival, and pay automatically through the app with a credit card. The company markets its clear pricing, convenient charging to credit cards, and easy and transparent use. Both driver and rider can also provide a rating on the experience—a prime feature of sharing services in general. UberX, the cheapest of its options, is roughly 40 percent lower in price than the average taxi, although Uber does utilize surge pricing, which raises the price of the ride during high-demand hours. Uber takes a 5 to 20 percent cut of the price of the ride and the rest goes into the driver's account. Drivers have great flexibility in their schedules and use their own vehicles to provide rides, and anyone can drive for Uber with a few requirements regarding background checks, vehicle ownership, and car insurance. There are other benefits as well. Some studies have looked at the impact ride sharing has on drunk driving rates: one study found that between 2009 and 2014, UberX resulted in a decrease in drinking and driving–related deaths of between 3.6 and 5.6 percent in Californian cities where the service is offered (Stevenson 2015).

One of the challenges provided by Uber is how ride sharing should be regulated. The largest voice against Uber has come from the taxicab industry—which had been highly regulated in U.S. cities. In New York City, regulation of cabs that are allowed to respond to "street hails" is based on a medallion system; the taxi medallion itself is a small metal plaque that is affixed to a vehicle, and it is illegal to operate certain types of taxis without one. The city controls the number of medallions. Some wonder whether Uber drivers should be subject to the same regulations that taxi drivers are subject to. Taxi drivers must meet licensing and training requirements, must undergo background checks, have licenses reviewed, and must regularly have cab inspections. Taxi drivers have complained that Uber has taken away business by offering the same services but with lax regulation. The argument is that by not owning taxi medallions, Uber skirts compliance costs, dilutes the market, and unfairly competes with taxis (Posen 2016). And according to a working paper from the New York University Marron Institute of Urban Management, "sharing firms have proven remarkably resistant to regulatory pushes to limit their growth, displaying uncanny abilities to rally consumers as political advocates" (Rauch and Schleicher 2015, 3).

How Are Cities Responding?

Some cities have attempted to ban Uber and similar ride-sharing services. Cities from Vancouver to Rio de Janeiro consider Uber illegal, and even whole countries such as Italy and Spain have banned the service (Goodyear 2015). Rio de Janeiro was the first city in Brazil to announce such a ban. Mayor Eduardo Paes signed legislation in September 2015 banning Uber and similar technologies, and drivers can receive fines of nearly $500 (AP 2015). However, Uber continues to operate in Rio under a court injunction.

Vancouver in British Columbia, Canada, seems like a likely place for Uber to flourish, considering it has a thriving tech scene that employs thousands of young people. However, British Columbia's Passenger Transportation Board (PTB), a six-member board that sets fares and rules for the taxi, black car, and limousine industries, categorizes Uber as a limousine service (De Vynck 2016). The PTB told Uber that it must follow the rules of limo services, which means setting a minimum fare of $75 in order to keep them from competing with taxis. In the view of the PTB, Uber was a limousine company breaking the law. Uber left Vancouver 6 months after it began operating there.

Uber began operating in San Antonio, Texas, in April 2014 but then threatened to close operations in the city 8 months later because legislators wanted Uber drivers to submit to fingerprinting and random drug tests—requirements that Uber thought was too burdensome for drivers. Uber and the city were engaged in confidential negotiations, and the city thought they were near a deal when Uber made good on its threat to exit in April 2015 (Griswold 2015). The company stated in a blog post that: "City officials have created a regulatory climate that makes it impossible for us to meet the high standard of service that riders from over 170 cities across the U.S. have come to expect" (Uber Newsroom 2015). In the end, Uber returned to the city after the city agreed to Uber's terms. A 10-fingerprint background check is now optional for drivers operating through ride-share apps, although it is required for taxis and pedi-cabs. One article in *Slate* called Uber "a company accustomed to steamrolling local governments" and one that plays "a brash brand of politics" (Griswold 2015).

But other cities are embracing the ride-share model, recognizing that Uber and similar companies offer a service that people want and increasingly demand. Washington, D.C., passed legislation embracing ride sharing: its city council passed the Vehicle-for-Hire Innovation Act (VIAA) in 2014, which does not require ride-share services to comply with the same licensing requirements as taxicabs. It instead sets a minimum insurance requirement and background check, registration, and operating requirements (Posen 2016). The act essentially legalizes Uber in Washington, D.C., and gives the District's Taxicab Commission the task of enforcing regulations (Aratani 2014).

Responses by city governments are far from uniform. In San Francisco, according to the San Francisco Municipal Transportation Agency (SFMTA), within 18 months of Uber's introduction there was a 65 percent decline in taxicab use. The California Public Utilities Commission is the regulatory body that oversees ride-sharing services in the state of California. The state adopted ride-share regulations in 2013 when it coined the term "transportation network company" (TNC) to create a new class of companies that provide transportation services but don't technically own cars or employ drivers (Goodyear 2015). While the state regulates ride-sharing networks, municipalities in California regulate taxis.

One city where Uber is highly regulated is New York City, where Uber has caused the price of taxi medallions to decline: in 2014 the average price was 17 percent below its peak in 2013 (Bond 2015, 90). The Taxi & Limousine Commission (TLC) has regulated the taxi industry in New York City since the 1970s (Posen 2016). Ride-sharing drivers in New York City have to gain TLC commercial licenses, which is the only U.S. city that requires this step. Uber drivers must have an eligible vehicle with TLC plates, must provide a social security card, a state driver's license, Department of Motor Vehicle registration, insurance card, and certificate (Kunkle 2016). According to a report by the city of New York: "In New York City, these companies face added requirements under the TLC's for-hire vehicle legal and regulatory framework; drivers are professionals and must hold a commercial license, as well as undergo more rigorous background checks and training. Though drivers may operate their personal vehicles, those vehicles must submit to City inspection" (City of New York 2016b, 3). Taxis have a $0.50 surcharge per ride that supports the Metropolitan Transportation Authority (MTA), but instead ride-share passengers pay 8.875 percent sales tax, of which 4.5 percent goes to the city and 0.375 percent goes to the MTA (Meyer 2016). Although Uber has so far complied with city regulations, in summer 2015 New York City Mayor Bill de Blasio proposed setting a cap on the number of for-hire cars operating in the city, citing congestion and pollution issues. Yellow and green cab fleet sizes are capped at roughly 15,237 and 18,000, respectively, while livery, black car, and limousines, including e-dispatch ride shares, have no such restrictions. The city estimates that there are roughly 148,000 licensed for-hire vehicles in New York City (City of New York 2016b, 8). This led to a study on the impact that ride-sharing companies have on the city, which was completed by McKinsey for roughly $2 million and published in early 2016 (Meyer 2016). The report found that the decreased traffic speeds in Manhattan could not be attributed to ride sharing. The study also found that ride sharing has increased transportation options for low-income, outer-borough New Yorkers. Areas such as Astoria, Harlem, Jackson Heights, and Washington Heights experienced more than

1,200 percent growth in monthly UberX rides throughout the course of 2014 (Meyer 2016).

Other Issues

Uber reported $1.27 billion loss in the first half of 2016, and the company attributes these losses to driver subsidies, especially in China (Somaney 2016). In order to build a strong customer base when Uber enters a new city, the company provides subsidies to drivers to offer promotions, such as discounts and free rides. Uber was spending about $1 billion a year in China to build its customer base through these promotions, but this was holding up its stock offering as well as expansion into other global markets. Uber has been profitable in the United States and Canada, but the substantial losses in 2016 underscore the difficulties faced when companies try to build and expand. Uber didn't face regulatory challenges in China, but when it entered the Chinese market in 2014, there was already a strong ride-sharing market dominated by two Chinese companies: Didi Dache and Kuaidi Dache. In 2014, Didi was the largest ride-hailing service in China, averaging 5.22 million ride requests per day (Yang, Su, and Fang 2016). In February 2015, Didi and Kuaidi merged to form one company: Didi Chuxing. Uber spent the first 2 years in the Chinese market fighting with these companies and in August 2016 agreed to sell its subsidiary company, Uber China, to Didi Chuxing (Isaac 2016).

Uber has also introduced some new programs that might actually deter people from using public transit. For example, in New York City, Uber began offering UberPool rides for a $5 flat rate, valid only on weekdays between 7 a.m. to 10 a.m. and 5 p.m. to 8 p.m. in Manhattan below 125th Street. In addition, Uber offered a promotion with Gilt City (a lifestyle website that sells discount offers for experiences) for "Commute Cards" that grant unlimited rides during rush hour for a flat rate; $49 for 2 weeks, $79 for 4 weeks, or $159 for 8 weeks. Both of these options essentially make commuting via car cheaper than taking the New York City subway, assuming riders are using an unlimited subway pass twice a day on weekdays only.

While Uber continues to operate and expand, a number of questions remain. Should drivers be considered independent contractors or employees? Should they be allowed to form unions? How much insurance coverage should be required? Should sales tax revenue they generate go toward mass transit? These are the types of questions and issues that ride-sharing operations have generated.

HOME SHARING: AIRBNB

Airbnb, the largest and most well-known home-sharing platform, is an online marketplace that for a small processing fee connects people with spare rooms or vacant homes to those seeking accommodations. Since it was founded in 2008, the San Francisco–based company has served more than 60 million guests and become practically synonymous with the sharing economy. In late 2007, founders Brian Chesky and Joe Gebbia were having trouble paying their high rent in San Francisco so they put up a website, Airbnb.com, to advertise their floor space for a small fee. After hosting several people in the first week, Chesky and Gebbia realized the business potential and decided to build a bigger site with more listings. In 2009, the company received $600,000 in seed funding, allowing it to increase its offerings and expand internationally. By the end of 2011, Airbnb had more than 2 million total nights booked around the world. Today, the company operates in more than 34,000 cities and 191 countries. It is reportedly valued at $25 billion, more than most traditional hotel chains, including Hyatt and Wyndham (Airbnb 2016; Geron 2013).

Airbnb labels itself as the "world's leading community-driven hospitality company" (Airbnb 2016, 2). Its popularity stems from its ease of access, customer-centric approach, and affordability. The user-friendly site allows people around the world to connect with little more than a few taps of their smartphones. The platform's searchable marketplace and transparent availability calendar are easily accessible and give travelers a wide range of lodging options, often at much lower costs than that of hotels in the same neighborhood (Jonas 2016). Unlike traditional businesses, the founders didn't have to build any infrastructure—it already existed in the world's cities. All the company needed was the Internet. This allowed Airbnb to very quickly build up a greater choice of rooms in terms of location, price, and amenities. With more than 2 million listings worldwide, Airbnb hosts offer almost every type of overnight accommodation imaginable: apartments, condominiums, houses, inns, mansions, castles, tents, teepees, and even yachts. In addition, guests get to enjoy the experience of living like a local and staying in the heart of the city (McNichol 2015).

Airbnb has helped pioneer a market that is radically changing consumer hospitality preferences and behaviors (PwC 2015, 23). Founders Chesky and Gebbia wanted to be entrepreneurs, but neither wanted to "create more stuff that ends up in a landfill" (Kirkland 2014). They created a company that at its core is based on sharing, just as the sharing economy itself was proliferating. According to Chesky, "We never considered the notion we were participating in a new economy. We were just trying to solve our own problem. After we solved our own problem, we realized many other people want this" (Geron 2013). For some, the appeal of Airbnb is that it embraces a sense of community and offers the opportunity to interact with others. The site allows travelers to connect with local hosts and receive travel tips in a personalized fashion. Some contend that Airbnb is helping to close critical gaps in the sharing hospitality industry; namely, issues of friction and trust (PwC 2015, 23). Airbnb is also perceived to be more environmentally sustainable. An Airbnb-funded study conducted by Cleantech Group (CTG) found that traveling on Airbnb results in significant reduction in energy and water use, greenhouse gas emissions, and waste and encourages more sustainable practices among both hosts and guests (Airbnb 2014). However, several environmental advocates questioned the study's findings, accusing the company of greenwashing. I think that if the result of Airbnb is that fewer new hotel rooms are constructed, it might have a positive impact on the environment, but that would depend in part on the carbon and environmental impact of the home now being more fully utilized.

Similar to Uber, a major issue with Airbnb is how it should be regulated. Until recently, Airbnb operated without much oversight, and its hosts were not subject to traditional hotel regulation (Chafkin and Newcomer 2016). However, the rapid rise of Airbnb has sparked a public debate in cities worldwide about the real-world consequences of this online marketplace. Proponents claim that in many cities, the surging popularity of Airbnb is bringing tourists into neighborhoods they never visited before, creating new economic opportunities for residents and local businesses

(Anderson 2016). They also argue that Airbnb promotes better resource allocation and contributes to smart, sustainable cities. Critics, in contrast, claim it is fueling a lucrative underground economy, promoting the evasion of taxes, and contributing to an acute shortage of affordable housing (Herman 2014). Some also see a threat to safety, affordability, and residential character of local communities (Jonas 2016, 221). The biggest problem for Airbnb as it expands stems from landlords pressuring tenants to move out so they can turn their apartments into illegal hotel units. These concerns have prompted local and state governments to try to regulate Airbnb's services.

How Are Cities Responding?

The primary concern of many cities is that Airbnb and its hosts are reducing tax revenue and operating illegally, and a growing number of cities are taking action to regulate Airbnb's operations (Herman 2014). The stiffest opposition to Airbnb has come from the cities of New York and San Francisco. In 2014, New York Attorney General Eric Schneiderman filed a suit against Airbnb for refusing to provide the names of its New York hosts, some of whom regulators suspected may not have paid taxes on income and may be violating New York housing laws (Harpaz 2016). It is illegal in New York to rent out an apartment for less than 30 days without the occupants present. Airbnb fought back, claiming protection of customer privacy, but lost the ruling and was forced to submit the data. The attorney general's office analyzed Airbnb's bookings from 2010 to 2014 and found that as many as 72 percent of the Airbnb listings in the city—totaling around 36,000 at the time—might be illegal. The report also found that property owners using the site evaded millions of dollars in annual hotel taxes (Dingman 2015; New York State Office of the Attorney General 2014). In June 2016, New York State legislators passed a law that would subject Airbnb and other short-term rental site hosts to heavy fines for allowing rentals that violate the state's short-term rental laws (Clampet 2016).

Airbnb's hometown, San Francisco, has also struggled to regulate the growing home-sharing platform. San Francisco's policy is similar to New York's: Airbnb rentals are allowed only if hosts are full-time residents and rentals are capped at 90 days. However, it is estimated that more than a quarter of San Francisco's Airbnb properties listed on the site violate the city's 90-day annual limit on such rentals. In February 2015, San Francisco enacted a new rule requiring Airbnb hosts to register with the city, but a vast majority have not complied. As of March 2016, only about a fifth of the estimated 7,000 hosts had registered their properties (Said 2016). Now, the city is holding Airbnb responsible for its hosts and will impose a fine on the company of $1,000 per day for each unregistered listing that the city discovers (Slee 2016).

Cities in the United States and around the world are trying to figure out how to regulate home-sharing services. In 2016, Chicago enacted a house rental law that sets various kinds of limits for short-term rentals (Byrne 2016). Housing officials in Paris, one of Airbnb's most popular vacation destinations with an estimated 40,000 listings, began cracking down on secondary apartments set up specifically as short-term rental units, with officials fining violators up to $30,000 each (Schechner 2015). Iceland recently began to require registration for all hosts. In 2015, Barcelona became the first city to penalize Web platforms, including Airbnb, for advertising illegal "unlicensed tourist accommodations" (Garcia 2016). Beginning May 1, 2016, Berlin banned landlords from renting out apartments to short-term visitors, with only a few exceptions. The penalty for breaking the law can be as high as a $100,000 fine on the hosts (O'Sullivan 2016). Amsterdam and London, however, have been far more receptive to Airbnb than other European destinations. In February 2014, Amsterdam became the first city to pass so-called Airbnb-friendly legislation, a cooperative effort in which the city will levy a tourist tax on rentals, while Airbnb will work to ensure potential hosts are aware of all pertinent rules and regulations (Van Daalen 2014). And Londoners interested in renting out their properties on Airbnb have benefited from a new amendment to the city's housing legislation passed in Parliament in March 2015, which allows homeowners to rent out their house, flat, or spare rooms for up to 3 months a year.

Airbnb is not pleased with the new regulations. Airbnb maintains that it has a positive impact on cities and continues to present its business as low-impact, made up of everyday hosts occasionally renting out their own homes (Slee 2016). The company also argues that it's not responsible for hosts that don't comply with the law. However, as more and more cities begin to tighten regulations, Airbnb is trying to work more with city governments. "We are less confrontational, and we try to more pro-actively open a conversation with cities," says co-founder Nathan Blecharczyk (Dingman 2015). In 2015, Airbnb launched the Community Compact, a document based on the company's core principles and informed by the lessons they've learned about how best to partner with cities (Airbnb 2016, 2). Airbnb announced it would start to investigate and remove people running illegal hotels in New York and San Francisco. In Europe, which makes up 70 percent of its revenue, the company has tried to woo government officials, offering to pay tourism taxes, encouraging new laws that bless its activities, and commissioning studies showing it boosts local economies. In 2016, it detailed new measures to encourage hosts in Europe to pay income taxes on rental revenue (Byrne 2016).

Recognizing that their own pace of innovation has moved faster than the pace of regulation, Airbnb has begun lobbying to change those laws so that the revenue generated through their peer-to-peer home-sharing model is taxed appropriately. So far, a few early adopters, including San Francisco, Portland, Ontario, Alabama, and Florida, have taken the lead on making deals with Airbnb and are now collecting tax revenue from the company on behalf of property owners. Airbnb has also reached an agreement with Paris on collecting a tourist tax on behalf of the city. Others have been less enthusiastic. The Santa Barbara City Council banned Airbnb-style short-term vacation rentals and, in Virginia, proposals promoting home sharing were postponed largely due to strong opposition from hotel groups (Harpaz 2014).

In 2016, a coalition of 10 cities across the globe came together to collaborate on a framework that will provide these companies with ground rules to abide by. The coalition, which includes New York, San Francisco, Paris, Seoul, Athens, Barcelona, Amsterdam, and Toronto, are jointly preparing a unified "rule book" in an effort to leverage their combined size to promote clearer ground rules. The move will help to normalize the experience of both users and businesses who are frequently caught up in the hodge-podge of current regulations. While there's no policy drawn up yet, multiple city representatives met for the first time in May 2016 to discuss the issue (Mawad, Fouquet, Goldman 2016).

Other Issues

Other problems that city governments and state regulators foresee with Airbnb include potential influxes of travelers who will transform residential neighborhoods into hotel districts and a current lack of oversight and accountability over Airbnb-related adherence to zoning laws. Affordable housing advocates have now joined the conversation, saying that sites like Airbnb are contributing to the crisis.

Airbnb says it contributes to local economies, in part, by attracting travelers who might not stay at hotels. According to an Airbnb report, "over 75 percent of Airbnb listings are outside main hotel districts, and typically up to 50 percent of Airbnb guest spending occurs in neighborhoods where they stay" (Airbnb 2016, 4). It also claims that its service is adding billions to spending in cities, with its own figures purportedly showing an economic impact of roughly $500 million in New York City, $300 million in London, and $150 million in Barcelona alone (Herman 2014). However, in some of these neighborhoods, there can be an unfortunate side effect—an influx of tourists, and the money they bring, can lead to higher rents and retail prices and the displacement of locals who can no longer afford to live there (Anderson 2016). In cities like Vancouver, London, and New York, sites like Airbnb are diverting rental stock that could have been rented to local, long-term tenants instead (Anderson 2016).

One concern with Airbnb is its effect on local housing prices. In high-rent cities like New York and San Francisco, Airbnb is often portrayed as helping cash-strapped individuals bring in extra income to afford their rent. However, in the long term, some believe it will likely exacerbate the city's housing crisis by allowing landlords to charge more in rent because their tenants can turn to this secondary market to make up the difference (Herman 2014). While Airbnb makes claims to the contrary, two important studies have found that hosts who rent a single room in their home are no longer the norm. One study of 90,000 hosts in 18 cities found that 40 percent had multiple listings, while another one commissioned by the *San Francisco Chronicle* newspaper found that nearly 5,000 listings in that city were for entire apartments or houses (Herman 2014). Affordable housing advocates have joined forces with the hotel industry and labor unions to demand tighter regulation. Airbnb says it is aware of the concerns of residents in neighborhoods with high amounts of hosts and is committed to working with cities and communities to prevent this. The company also argues that there are many other factors at play, such as long-standing city policies around housing (Said 2016).

The company has also had to deal with discrimination issues. Studies by Harvard University researchers released in 2014 and 2015 suggest that black Airbnb hosts make less money than white hosts and that black users are sometimes refused service on the basis of race (Chafkin and Newcomer 2016).

Thanks to Airbnb, the concept of home sharing is now well entrenched in urban life. The challenge now is how cities effectively regulate its services in a way that benefits a wide array of interest groups—from Airbnb's loyal customer base, to neighborhood residents, to affordable housing advocates, to hotel groups, and more. Policy makers and tech leaders need to work together to craft enforceable regulations that protect public interest as well as innovation. With modern technologies, the intersection of urban space and cyberspace provides an unsurpassed platform for a more inclusive and environmentally efficient sharing economy.

Conclusion

The sharing economy has the potential to transform both business and cities in unprecedented ways and to boost urban resilience and sustainability in the process. It is a new way of consuming: using without owning. The car you drive, or are driven in, is not your own. The place you visit is someone else's property. All members of the sharing economy—participants, entrepreneurs, policy makers, and other leaders—have an essential role to play in making this happen (Chang 2014). According to Rachel Botsman in a *Harvard Business Review* article: "The real power of the collaborative economy is that it can serve as a zoom lens, offering a transformative perspective on the social, environmental, and economic value that can be created from any of a number of assets in ways and on a scale that did not exist before. In that transformation lie threats—and great opportunities" (Botsman and Rogers 2010). Rachel Hatch, research director at the Institute for the Future, a nonprofit based in Palo Alto, California, and a regular Airbnb host herself, adds: "If I were to forecast it for the next decade, cities that will thrive will be the ones that pause to consider how to leverage their sharing economy, while preventing gentrification and leveraging the city's character" (Anderson 2016).

In a global economy, a distinct sense of place is difficult to maintain. If the hotel you stay in due to a loyalty club looks the same on every continent, you can easily forget where you are. However, if you stay in someone's home, each dwelling has some element of local to offer. While the America-first ideology we sometimes see in American politics is attempting to push back globalization, the technological and economic force of globalism will be too strong to resist. Rather than resist the tide, the sharing economy provides a way to ride with it: using the technology that brought us a global economy in order to borrow and use the distinct local elements that an individual prefers.

ANALYSIS OF SUSTAINABLE URBAN LIVING AND THE SHARING ECONOMY: UBER AND AIRBNB

Sustainable Living as an Issue of Values

The sustainable city, its infrastructure, and impact on the planet is in many ways governed by a value construct of people attracted to living in the city. The development of the sharing economy is a major social and cultural phenomenon that should not be underestimated. The connection between consumption, possession, and ownership appears to be loosening. This is especially true of young people. A growing number see resources such as autos as a means toward an end rather than an end itself. They don't want to parade their car around town, but instead post photos of themselves on social media standing in iconic locations. It is a true manifestation of a post-scarcity economy when people can assume they can make use of anything they want whenever they want, without owning it. This changing relationship of people to property is a major paradigm shift in the definition of property.

The value of experience and of social interaction seems to be replacing, for some, the need to own the possessions that facilitate social interaction. Ownership of the latest technology remains important, but even with technology the need to possess the latest iPhone seems to be gradually replaced by the need for the latest software application. It is not that material goods are unimportant, but that they are less important than ideas, events, and experiences.

In addition, polling data indicate that a concern for environmental quality is higher among young people than old people, and strong majorities favor environmental protection. This cuts across ideology and political party and seems to be a function of the sociology of the millennial generation. A concern for healthy food, physical fitness, and wellness is closely associated with positive views of environmental protection. This serves to reinforce the sharing economy by viewing ownership of goods as less essential than access to products for use.

Sustainable Living as a Political Issue

An interesting aspect of sustainable living is that it is a cultural and social phenomenon and largely apolitical. While traditional environmentalism is sometimes associated with liberal politics, and certainly climate change policy appears to be, the sharing economy, renewable energy, energy efficiency, physical fitness, and wellness appear to cut across ideological divides. To the extent there is a political element, it appears to be to keep government away from the sharing economy. Efforts to regulate Uber and Airbnb are resisted, but not so much out of general resistance to regulation, but stemming from a perception that established interest groups and regulators are either misinformed or simply fighting innovation to protect the old economy and the benefits they derive from it.

Sustainable Living as an Issue of Science and Technology

Without GPS, inexpensive computing, cell phones, bar codes, credit card encryption chips, and a host of other technological innovations, the sharing economy would not be possible. To a considerable extent, the new economy is built on technology and is designed to share its benefits. Social media and free videography and photos allow experiences to be widely and instantly shared at no cost to the viewer or the producer.

In a more fundamental sense, the ability to assume that food, clothing, and shelter need not be a focus of our daily attention has been made possible by a wide range of labor-saving technologies from the internal combustion engine, to the tractor, to the electrical grid. Our lifestyle requires energy and technology, which has been used to replace human and animal labor in production. Our knowledge of medical science and the growth of medical technology have enabled wellness and physical fitness to be part of our daily routine.

Sustainable Living as an Issue of Public Policy Design

The sharing economy requires a new form of regulation that allows the public to be protected while limiting restrictions on those leasing their autos, homes, tools, or skills to that same public. The traditional command and control regulation and licensing is not an effective policy design, and so a new form of governance must be developed. Some elements of sustainable living do not require new forms of organization and living, but where the sharing economy is involved, a new governance design and structure will be required.

What has been termed the "gig economy," or short-term work assignments, also seems to be a part of this new lifestyle. People trade off freedom during the workday against job security and employer-funded benefits—particularly for health care, vacation, and retirement. Workers sacrifice the protection of unions and large corporations for freedom of movement. The social safety net will need to be redesigned to protect workers under these conditions. While this is not a prerequisite to the establishment of the new economy, it makes sense to anticipate these problems we are certain to see down the road. It will be interesting to see if the young adherents of the sharing economy change as they age and live in an aging society. Will they become more concerned about security, when mortality and illness become more relevant?

Sustainable Living as a Management Issue

The new organizational routines and incentive systems required for the sharing and gig economy are still under development. We do not know how to build and maintain these organizations. We are unlikely to see deep vertically integrated organizational structures. We do not have enough experience to know if the financing of these organizations and practices will provide an adequate return on equity. We also do not know if talented management can be attracted and retained by future Ubers and Airbnbs. We certainly can assume there will be a great deal of learning over the coming decades as some business and organizational models are abandoned.

Sustainable Living as a Multidimensional Issue

While sustainable living is best understood by examining it against a wide array of conceptual lenses, the most fundamental is to view it as an issue of values. Throughout the history of Western culture, possession of property has been a fundamental value, nearly always equated with the use of property. Separating use from ownership strikes me as a fundamental shift of values and of central importance to understanding an economy that maintains consumption while reducing the use of natural resources. While clearly technology is required to enable this change, people must value the experience of using material over possessing that good. That is central to the sharing economy and the element of sustainable living that it engenders.

CONCLUSIONS

11

TOWARD A SUSTAINABLE CITY

Human history is in many ways a story of the evolution of our social interaction and the development of the technology that affects our daily lives. How we behave toward each other reflects our values and ethics, and they have evolved as we have learned to meet our basic and then other than basic needs. Over the past two centuries, the pace of technological change has increased dramatically influencing every aspect of work and home life. Climate control, refrigeration, telephones, radio, electric illumination, televisions, autos, jets, computers, the Internet, smartphones, search engines, industrial agriculture—the list is virtually endless, and its impact on how we live, how we work, and how we spend our time is nearly impossible to calculate and understand.

In the course of transforming how we live and how we spend our time, we have gained greater mastery over our ability to affect our immediate environment. At one time, most people spent most of their time struggling to obtain food, clothing, shelter, water, and safety. In the modern world, those resources are assumed. Despite America's obsession with guns, most people do not hunt and do not need to defend themselves against dangerous enemies. Most people spend a relatively small amount of time and effort to feed and house themselves. At one time, most people spent most of their time securing the necessities of life. That is no longer the case for billions of people, and for those who do continue this struggle, in the modern world we consider them to be poor or extremely poor.

But in the process of building this technological marvel, a world unimaginable to people born at the start of the twentieth century or earlier, we have damaged natural systems ranging from groundwater to climate. We have introduced human-made chemicals into ecosystems that do not have the ability to absorb them without being damaged. These ecosystems produce the air, water, and food that our species requires to stay alive. With the human population of the world exceeding seven billion and likely to peak at about nine billion, we need to figure out how to manage the impact of our species on the planet. We need to do this while increasing the output of our high-throughput economy to ensure that poverty is ended and a global social safety net is put in place. This must be built on an understanding of the importance of the market economy and of individual reward and sanction for individual behavior. A regulatory and management system must permit individual accomplishment while preventing environmentally destructive organizational and individual behavior.

This requires a deep understanding of earth systems science and the impact of human-built technological creations and systems on ecosystems and on the global biosphere. To achieve this we need to:

- fund and pay for earth systems science and impact projections/analyses;
- develop sophisticated public-private partnerships to make the transition from a one-time, finite resource–based, throw-away economy to a renewable resource–based sustainable economy; and
- concentrate as much human population as we can into densely settled areas while ensuring those places are attractive, exciting, healthy, and productive places to live, learn, develop, work, and play.

The key to a sustainable, renewable resource–based economy is sustainable cities. These are places that attract people and achieve economies of scale in all of the urban systems summarized in chapter 2:

Energy
Water
Solid waste
Sewage treatment
Food
Public space
Transportation

This is not to prevent people from living in rural areas or to discourage wealthy people from having both city and country homes. People should live where they want to live. The goal is to make the city so attractive that people are drawn to them. In fact, to maintain the value of environmental protection, urban dwellers should be encouraged to experience nature on a regular basis. Ecotourism, school trips, vacations in the country, and sports like hiking, skiing, fishing, and hunting all have a place in building the value of environmental protection. But we want to avoid all nine billion of us spreading out in millions of low-density communities. Protecting the environment from global suburban sprawl might be possible someday, but not today.

The purpose of this book is to explicitly draw the connection between sustainable processes and sustainable places. Sustainable economic processes require sustainable local places. To do this, we must develop a deeper understanding of human technologies and the impact of those technologies on the living systems of living places.

Sustainable cities will not emerge all at once or without false starts. We need to learn how to create and maintain these places. It seems obvious, but worth repeating, that these cities must be built to attract residents and visitors. They need to deliver high quality of life and a sense of place and space that makes them unique and distinctive. Every city will have some sort of trade-off. Some will have better weather, some will have better museums, others will have better bars, and others will have better health care. But all should do everything they can to reduce the impact of their population on the quality of ecosystems and the air, water, and food that we require. The case studies that I provide in this book are designed to illuminate both the difficulty and the feasibility of building sustainable cities. The change is coming, but that path is far from direct.

I devoted a chapter of this book to the sustainable lifestyle because of my growing recognition of the relationship of social change to economic change and political change. The social or cultural move toward a new type of economic consumption makes the sustainable city not only possible, but necessary and probable. Young people want some of the things their parents have but want other things as well. They want experiences as much as they want possessions. They want to share and communicate those experiences with others. The growth of the World Wide Web and the lower price of communication and computing have changed the nature of information sharing and of social interaction. Your circle of friends is no longer limited to those you live in close proximity to, but the extent of electronic social interaction has prompted increased demand for live interaction. We do not know what shape that will take, but we know that nearly all live human interaction takes place in cities.

The social, economic, and technological changes we have experienced in the United States over the past half-century provide the foundation for the changes that are now under way. Feminism, the move to eliminate racism, changing norms of child rearing (being a parent used to be a status, parenting is now a verb), gay rights and marriage equality, physical fitness, wellness, enhanced psychological and physical health care—the list goes on, but we live in a different world than the one I grew up in. It makes it possible to think about quality of life in different ways. People care deeply about the food they eat, the air they breathe, and the water they drink. That often leads to a path of advocating environmental protection and trying to live a sustainable, less resource consumptive lifestyle.

Communication, education, social interaction, culture, art, music, theater, comedy, and the media allow you to observe, interact, and reflect. They do not require ownership. You need not buy the painting to enjoy it. Your entertainment can be streamed rather than recorded on a disc and played back. Your books can be stored in the cloud rather than on a case in your living room. The connection between the mind and physical objects has been reduced, as has the relationship between quality of life and material possessions. We require food, water, clothing, shelter, and other material comforts, but those items are increasingly assumed commodities and while we may be fooling ourselves, we spend less and less time pursing those items and more and more time acquiring and digesting information and ideas.

We now live in a globally interconnected economy and communication system. We are more interdependent than ever. As these connections have grown, we have also noticed that in many nations income inequality has also grown. I am personally less concerned with what the rich have than with what the poor lack. Food, clothing, and shelter are obvious needs, but in the emerging brain–based economy access to education and information/communication technology is nearly as important. The modern version of giving someone a fishing rod instead of a fish is to provide a young person with the opportunity to enter the global conversation. Long-term political stability requires that the very wealthy give up more of their income to ensure the health, education, and welfare of people who for one reason or another end up with less. We do not require income equality, but access to opportunity. The alternative is more terrorism, more incarceration, and more instability.

Cities have the advantage of density, and it is difficult for the very rich to completely avoid the very poor. It also reduces the cost of access. New York City is setting up wireless kiosks on many corners to allow people to connect to the Web for free and to recharge their electronic devices. Libraries have resumed their historic role as entry points for people needing information and skills. Instead of finding what they need in stacks of books, people sit at tables of computers, accessing the Web and printing cover letters and résumés.

There are many old business interests fighting to maintain their businesses even though they pollute and are ready for replacement. They will not go away, but time is not on their side. If the past two centuries tell us anything, it is that the tide of technological change cannot be stopped. It can be shaped and directed, but humans are too creative and the benefits of technology are too seductive. We certainly have traded off many benefits to get the comforts and lifestyle we have. We have sacrificed privacy, self-sufficiency, and freedom of movement. These changes have been gradual enough that many people don't realize they've taken place and cling to a politics and value system that bears little resemblance to how they live or what they need to defend.

For the past 40 years I have maintained my interest in how average people can make meaningful policy choices about complex technologies they do not understand. They know that a chemical covering makes their pots and pans easier to clean, but they don't realize that the process of producing that coating could be quite dangerous if poorly managed. There are so many technologies and so many costs and benefits, how does one know what to focus on and what can be ignored? How do we link people to the critical decisions that will determine how, where, and even if they will live? My answer always comes down to improved literacy on the part of those receiving messages and better communication skills on the part of those sending the messages.

But that answer doesn't really provide a way to manage the growing complexity and toxicity of the world. It is interesting to note that despite the growing dangers, people are living longer and healthier lives. A great deal of the technological might and human ingenuity of our culture has been focused on medicine, which attempts to treat the illnesses caused by both natural and human causes. But that technology must be delivered in real time, and cities are where that takes place. If you are stricken ill, a modern communication system speeds help to your location, and a well-equipped modern ambulance has some of the same features you will find in a hospital emergency room. That is why many people survive shootings, heart attacks, and other sudden health traumas.

In the sustainable city, the communication and learning process can be a group process. Physical proximity makes it easier to gather a group to learn together and to act together. In a hospital, a team of experts can interact in an emergency room and stabilize a patient. Experts engage with groups of stakeholders who can help each other understand whatever is being discussed. It turns out that place is a really important tool to handle complexity. In addition to convening power, the public has access to trusted experts, such as a local pharmacist who can explain the interactive issues raised by a new prescription or a plumber who can help you find someone to test your pipes for contamination. It is also possible to view change across time and space as new buildings and parks go up and as new playgrounds fill up with families.

While there is no longer enough unspoiled nature for every child to have their own section of a local stream to call their own, it is possible to consciously give our children routine experiences in nature and build their appreciation of the environment that life depends on. The balance between human and natural has been decisively shifted toward human. Sustainable cities are needed to avoid a completely artificial world. We could survive in such an unnatural world, but my value system would prefer to prevent the destruction of the natural world. If we learn to protect it, we can have the excitement and comfort of the cities we love, alongside the natural world I believe we will always need.

WORKS CITED

A.T. Kearney. 2016. "Sustainability." https://www.atkearney.com/sustainability.

Abunimah, Ali. 2015. "Boycott-hit Veolia Dumps Jerusalem Rail, Completes Israel Withdrawal." *Electronic Intifada*, August 28. https://electronicintifada.net/blogs/ali-abunimah/boycott-hit-veolia-dumps-jerusalem-rail-completes-israel-withdrawal.

Accenture. 2016. "Why a Sustainable Supply Chain Is Good Business." https://www.accenture.com/us-en/insight-outlook-why-sustainable-supply-chain-is-good-business.

Adams, Mary. 2015. "Intangibles and Sustainability: Holistic Approaches to Measuring and Managing Value Creation." *Journal of Applied Corporate Finance* 27 (2): 87–94. doi:10.1111/jacf.12121.

Advameg City Data. 2008. "Cities of the World: Jerusalem." http://www.city-data.com/world-cities/Jerusalem-Economy.html.

Ahronot, Yediot. 2015. "Global Light Rail Award for Ministry of Transport and Jerusalem Municipality for the Light Rail." Israel Infrastructure Fund, October 25. http://www.iif.co.il/presscenter/global-light-rail-awards-for-ministry-of-transport-and-jerusalem-municipality-for-the-light-rail/.

Airbnb. 2014. "New Study Reveals A Greener Way to Travel: Airbnb Community Shows Environmental Benefits of Home Sharing." July 31. https://www.airbnb.com/press/news/new-study-reveals-a-greener-way-to-travel-airbnb-community-shows-environmental-benefits-of-home-sharing.

——. 2016. "Generating $2 Billion in Potential Tax Revenue for America's Cities." https://www.airbnbaction.com/wp-content/uploads/2016/01/USCM-Tax-Report.pdf.

Allman, Tammy. 2013. "Why Hong Kong Is Drowning in Its Own Waste with No Buoy In Sight." *Culturist*, July 9. http://www.thecultureist.com/2013/07/09/hong-kong-waste-problems/.

American Nutrition Association. 2015. "USDA Defines Food Deserts." http://americannutritionassociation.org/newsletter/usda-defines-food-deserts.

American Student Assistance. 2013. "Life Delayed: The Impact of Student Debt on the Daily Lives of Young Americans." http://www.asa.org/site/assets/files/3793/life_delayed.pdf.

Anderson, Brook. 2016. "How Airbnb Could Change Life for City Residents." *Wall Street Journal*, April 22. http://www.wsj.com/articles/how-airbnb-could-change-life -for-city-residents-1461550187.

Anderson, Monica. 2015. "Partisans Differ Sharply on Power Plant Emissions Limits, Climate Change." Pew Research Center, August 3. http://www.pewresearch.org /fact-tank/2015/08/03/partisans-differ-sharply-on-power-plant-emissions -limits-climate-change/.

Anderson, Monica. 2016. "For Earth Day, Here's How Americans View Environ- mental Issues." Pew Research Center, April 22. http://www.pewresearch.org/fact -tank/2016/04/22/for-earth-day-heres-how-americans-view-environmental-issues/.

AOML (Atlantic Oceanographic & Meteorological Laboratory). 2014. "Hurricane Research Division: The Thirty Costliest Hurricanes 1900-2013." National Oceanic & Atmospheric Administration. http://www.aoml.noaa.gov/hrd/tcfaq/costliesttable .html.

AON Benfield. 2012. "Annual Global Climate and Catastrophe Report." http:// thoughtleadership.aonbenfield.com/Documents/20130124_if_annual_global _climate_catastrophe_report.pdf.

——. 2013. "Annual Global Climate and Catastrophe Report." http://thoughtleadership. aonbenfield.com/Documents/20140113_ab_if_annual_climate_catastrophe_report .pdf.

——. 2014. "Annual Global Climate and Catastrophe Report." http://thoughtleadership .aonbenfield.com/Documents/20150113_ab_if_annual_climate_catastrophe_report .pdf.

——. 2015. "Annual Global Climate and Catastrophe Report." http://thoughtleadership .aonbenfield.com/Documents/20160113-ab-if-annual-climate-catastrophe-report .pdf.

AP (Associated Press). 2015. "Rio de Janeiro Becomes First City in Brazil to Ban Uber." *Guardian*, September 30. https://www.theguardian.com/world/2015/sep/30 /rio-de-janeiro-brazil-uber-ban.

Aratani, Lori. 2014. "D.C. Council Okays Bill to Legalize Lyft, Sidecar, UberX-Type Services in the District." *Washington Post*, October 28. https://www.washingtonpost .com/news/dr-gridlock/wp/2014/10/28/d-c-council-okays-bill-to-legalize-lyft -sidecar-uberx-type-services-in-the-district/.

Archdeacon, Kate. 2009. "Victor Civita Plaza: Living with a Site's History." Sustainable Cit- ies Net, October 9. http://www.sustainablecitiesnet.com/models/victor-civita-plaza/.

Athman, Julie, and Martha C. Monroe. 2004. "The Effects of Environment-Based Edu- cation on Students' Achievement Motivation." *Journal of Interpretation Research* 9 (1): 9–25.

Baer, Drake. 2014. "The Making of Tesla: Invention, Betrayal, and the Birth of the Roadster." *Business Insider*, November 11. http://www.businessinsider.com/tesla-the -origin-story-2014-10.

Balkan, Elizabeth. 2012. "The Dirty Truth About China's Incinerators." *Guardian,* July 4. http://www.theguardian.com/environment/2012/jul/04/dirty-truth-chinas -incinerators.

Batarce, Marco, Juan Carlos Muñoz, Juan de Dios Ortúzar, Sebastian Raveau, Carlos Mojica, and Ramiro Alberto Ríos. 2015. "Evaluation of Passenger Comfort in Bus Rapid Transit Systems." Inter-American Development Bank. https://publications. iadb.org/bitstream/handle/11319/6859/Evaluation%20of%20Passenger%20Comfort %20in%20Bus%20Rapid%20Transit%20Systems.pdf?sequence=1.

Bebon, Joseph. 2016. "U.S. Solar Job Creation Is 'Booming,' California Ranks No. 1." Solar Industry, February 10. http://solarindustrymag.com/u-s-solar-job-creation-is -booming-california-ranks-no-1.

Beketov, Mikhail A., Ben J. Kefford, Ralf B. Schäfer, and Matthias Liess. 2013. "Pesticides Reduce Regional Biodiversity of Stream Invertebrates." *Proceedings of the National Academy of Sciences of the United States of America* 110 (27): 11039–11043. doi:10.1073 /pnas.1305618110.

Belk, Russell. 2010. "Sharing." *Journal of Consumer Research* 5: 715–734. doi:10.1086 /612649.

Bennett, Michael, and David Warfield Teague, eds. 1999. *The Nature of Cities: Ecocriticism and Urban Environments.* Tucson: University of Arizona Press.

Berfield, Susan. 2015. "Inside Chipotle's Contamination Crisis." *Bloomberg Businessweek,* December 22. http://www.bloomberg.com/features/2015-chipotle-food-safety-crisis/.

Berg, Madeline. 2015. "The History of 'Death Avenue.'" Friends of the High Line, October 22. http://www.thehighline.org/blog/2015/10/22/the-history-of-death-avenue.

Berkeley Lab. 2016. "Microgrids: New York University." https://building-microgrid.lbl .gov/new-york-university.

Bigi, Alessandro, Elisabetta Corvi, and Gabrielle Ng. 2007. "The European Millennials vs. the U.S. Millennials: Similarities and Differences." Paper presented at Quarto Convegno Annuale della Societa Italiana Marketing. Presented in Rome, IT, 5-6 October 2007. http://www.unibs.it/sites/default/files/ricerca/allegati/Paper68.pdf.

Biophilic Cities. 2015. "Standards and Protocol for Participation in the Biophilic Cities Network." http://biophiliccities.org/wp-content/uploads/2015/03/BiophilicCities- NetworkProtocol.pdf.

BLS (Bureau of Labor Statistics). 2015a. "Employment by Major Industry Sector." United States Department of Labor. http://www.bls.gov/emp/ep_table_201.htm.

——. 2015b. "Employment Projections: 2014-24 Summary." United States Department of Labor. http://www.bls.gov/news.release/ecopro.nro.htm.

Bolster Creative. 2015. "The Millennial Shift." *2015 Workplace Trend Report.* http://www .bolstercreative.com/wp-content/uploads/2015/08/2015TrendReport_TheMillennial -Shift.pdf.

Bond, Andrew T. 2015. "An App for That: Local Governments and the Rise of the Sharing Economy." *Notre Dame Law Review Online*, 90 (2): 77–96.

Bonilla-Chacín, Maria Eugenia. 2013. *Promoting Healthy Living in Latin America and the Caribbean: Governance of Multisectoral Activities to Prevent Risk Factors for Noncommunicable Diseases.* Washington, DC: World Bank. doi:10.1596/978-1-4648-0016-0.

Boquet, Yves. 2011. "High-Speed Rail in China and Its Territorial Impacts." Presented at the 107th Annual Meeting of the Association of American Geographers (AAG). Presented in Seattle, WA, April 12. http://www.rechercheisidore.fr/search /resource/?uri=10670/1.7wxuba.

Botsman, Rachel, and Robert Rogers, 2010. *What's Mine Is Yours: The Rise of Collaborative Consumption.* New York: HarperCollins.

Bouwman, Lex, Kees Klein Goldewijk, Klaas W. Van Der Hoek, Arthur H. W. Beusen, Detlef P. Van Vuuren, Jaap Willems, Mariana C. Rufino, and Elke Stehfest. 2013. "Exploring Global Changes in Nitrogen and Phosphorus Cycles in Agriculture Induced by Livestock Production Over the 1900–2050 Period." *Proceedings of the National Academy of Sciences of the United States of America* 110 (52): 20882–20887. doi:10.1073/pnas.1012878108.

Bowen, Ted Smalley, and Adam Stepan. 2014. *Public Private Partnerships for Green Space in NYC.* New York: School of International and Public Affairs, Columbia University.

Boylan, Chris. 2016. "A Brief History of Tesla Cars in One Simple Infographic." *Clean Technica,* May 27. http://cleantechnica.com/2016/05/27/brief-history-tesla-cars-one -simple-infographic/.

Bradley, Ryan. 2016. "Tesla's Cheaper Model 3 Could Strain Charging Infrastructure." *MIT Technology Review,* March 29. https://www.technologyreview.com/s/601063 /teslas-cheaper-model-3-could-strain-charging-infrastructure/.

Bradsher, Keith. 2013. "Speedy Trains Transform China." *New York Times,* September 23. http://www.nytimes.com/2013/09/24/business/global/high-speed-train-system-is -huge-success-for-china.html?_r=1.

Broder, Jonathan M. 2012. "Deconstructing New York City's High Line Park: The How, Why and Wherefore." *Journal of Transportation Law, Logistics & Policy* 79 (3): 245–252.

Buch, Tanja, Silke Hamann, Annekatrin Niebuhr, and Anja Rossen. 2013. "What Makes Cities Attractive? The Determinants of Urban Labour Migration in Germany." *Urban Studies* 51 (9): 1960–1978. doi:10.1177/0042098013499796.

Bullis, Kevin. 2013. "How Tesla Is Driving Electric Car Innovation." *MIT Technology Review,* August 7. https://www.technologyreview.com/s/516961/how-tesla-is-driving -electric-car-innovation/.

Bullock, Richard, Ying Jin, Gerald Paul Ollivier, and Nanyan Zhou. 2014. *High-Speed Railways in China: A Look at Traffic.* World Bank Report, No. 93227. http:// documents.worldbank.org/curated/en/451551468241176543/High-speed-railways -in-China-a-look-at-traffic.

Bump, Philip. 2014. "Here Is When Each Generation Begins and Ends, According to Facts." *Atlantic,* March 25. http://www.theatlantic.com/national/archive/2014/03 /here-is-when-each-generation-begins-and-ends-according-to-facts/359589/.

——. 2015. "Climate Change Is More of a Priority for Voters than You Might Think." *Washington Post*, April 3. https://www.washingtonpost.com/news/the-fix/wp/2015/04/03/on-climate-change-people-who-want-action-have-more-passion-and-more-numbers/.

Burger, Andrew. 2015. "Renewable Energy Is Beginning to Power Africa." *Renewable Energy World*, May 21. http://www.renewableenergyworld.com/articles/2015/05/renewable-energy-is-beginning-to-power-africa.html.

Byrne, John. 2016. "Airbnb Rules Easily Pass Chicago City Council Despite Vocal Opposition." *Chicago Tribune*, July 22. http://www.chicagotribune.com/news/local/politics/ct-chicago-city-council-airbnb-rules-met-20160622-story.html.

Cain, Alasdair, Georges Darido, Michael R. Baltes, Pilar Rodriguez, and Johan C. Barrios. 2006. "Applicability of Bogotá's TransMilenio BRT System to the United States." Federal Transit Administration. https://nbrti.org/docs/pdf/Bogota%20Report_Final%20Report_May%202006.pdf.

Cannon, Sarah, and Lawrence Summers. 2014. "How Uber and the Sharing Economy Can Win Over Regulators." *Harvard Business Review*, October 13. https://hbr.org/2014/10/how-uber-and-the-sharing-economy-can-win-over-regulators.

Capek, Stella M. 2010. "Foregrounding Nature: An Invitation to Think About Shifting Nature-City Boundaries." *City & Community* 9 (2): 208–224. doi:10.1111/j.1540-6040.2010.01327.x

Capitol Riverfront BID. 2012. "Front Page News." http://www.capitolriverfront.org/_files/docs/caprivfpnfall2012.pdf.

——. 2016a. "About." http://www.capitolriverfront.org/about.

——. 2016b. "Canal Park History." http://www.capitolriverfront.org/canal-park/history.

CDC (Centers for Disease Control and Prevention). 2012. "A Look Inside Food Deserts." http://www.cdc.gov/features/FoodDeserts/.

——. 2013. "How Did Cause of Death Contribute to Racial Differences in Life Expectancy in the United States in 2010?" http://www.cdc.gov/nchs/data/databriefs/db125.htm.

CDP (Carbon Disclosure Project). 2015. "CDP Climate Change Report 2015: The Mainstreaming of Low-Carbon on Wall Street." https://www.cdp.net/CDPResults/CDP-USA-climate-change-report-2015.pdf.

Chafkin, Max, and Eric Newcomer. 2016. "Airbnb Now Has 100 Million Users and More Grown-Up Problems." *Skift*, July 11. https://skift.com/2016/07/11/airbnb-now-has-100-million-users-and-more-grown-up-problems/.

Champion, Tony. 2001. "Urbanization, Suburbanization, Counterurbanization and Reurbanization." In *Handbook of Urban Studies*, edited by Ronan Paddison, 143–162. London: Sage.

Chang, Alexandra. 2014. "Regulation Won't Kill the Sharing Economy. We Just Need New Rules." *Popular Science*, July 8. http://www.popsci.com/article/technology/regulation-wont-kill-sharing-economy-we-just-neednew-rules.

Cheshire, Paul C. 2006. "Resurgent Cities, Urban Myths and Policy Hubris: What We Need to Know." *Urban Studies* 43 (8): 1231–1246. doi:10.1080/00420980600775600.

Chiesura, Anna. 2003. "The Role of Urban Parks for the Sustainable City." *Landscape and Urban Planning* 68: 129–138.

Church, Wendy, and Laura Skelton. 2010. "Sustainability Education in K-12 Classrooms." *Journal of Sustainability Education* 1: 163–175.

Citizens Budget Commission. 2014. "12 Things New Yorkers Should Know About Their Garbage." http://www.cbcny.org/sites/default/files/REPORT_GarbageFacts_05222014.pdf.

City and County of San Francisco. 2012. "Mayor Lee Announces San Francisco Reaches 80 Percent Landfill Waste Diversion, Leads All Cities in North America." http://sfmayor .org/article/mayor-lee-announces-san-francisco-reaches-80-percent-landfill -waste-diversion-leads-all.

City of New York. 2007. *PlaNYC: A Greener, Greater New York*. http://www.nyc.gov /html/planyc2030/html/home/home.shtml.

——. 2014. "PlaNYC Progress Report 2014." http://www.nyc.gov/html/planyc2030 /downloads/pdf/140422_PlaNYCP-Report_FINAL_Web.pdf.

——. 2015. "Mayor's Management Report." Mayor's Office of Operations. http://www1 .nyc.gov/assets/operations/downloads/pdf/mmr2015/2015_mmr.pdf.

——. 2016a. "OneNYC 2016 Progress Report." http://www1.nyc.gov/html/onenyc /downloads/pdf/publications/OneNYC-2016-Progress-Report.pdf.

——. 2016b. "For-Hire Vehicle Transportation Study." Office of the Mayor. http://www1 .nyc.gov/assets/operations/downloads/pdf/For-Hire-Vehicle-Transportation-Study .pdf.

Clampet, Jason. 2016. "Airbnb Loses a Fight in New York as Legislature Passes Strict Advertising Law." *Skift*, June 17. https://skift.com/2016/06/17/airbnb-loses-political -battle-in-new-york-as-legislature-passes-advertising-law/.

Clarridge, Christine. 2001. "Cleaning Up Gas Works Park." *Seattle Times*, March 17. http://community.seattletimes.nwsource.com/archive/?date=20010317&slug=gaswo rks17m#_ga=1.49945190.1137150615.1472756411.

CNN Wire Staff. 2011a. "3 Nuclear Reactors Melted Down After Quake." *CNN*, June 7. http://www.cnn.com/2011/WORLD/asiapcf/06/06/japan.nuclear.meltdown/index .html.

——. 2011b. "Fresh Aftershocks in Japan Rouses Fear, Kills 2." *CNN*, April 8. http://www .cnn.com/2011/WORLD/asiapcf/04/07/japan.quake/index.html?hpt=T1&iref=BN1.

Cohen, Patricia. 2016. "The Cities on the Sunny Side of the American Economy." *New York Times*, March 31. http://www.nytimes.com/2016/04/01/business/economy/cities -where-us-economy-is-thriving.html.

Cohen, Steven. 2006. *Understanding Environmental Policy*. New York: Columbia University Press.

——. 2011. *Sustainability Management*. New York: Columbia University Press.

——. 2014. *Understanding Environmental Policy, Second Edition*. New York: Columbia University Press.

Conway, Hazel. 2000. "Parks and People: The Social Functions." In *The Regeneration of Public Parks*, edited by Jan Woudstra and Ken Fieldhouse, 9–20. London: Spon Press.

Cornell Waste Management Institute. 2001. "Final Report: Pay as You Throw for Large Municipalities." http://cwmi.css.cornell.edu/PAYTreport.pdf.

Cortright, Joe. 2014. "The Young and Restless and the Nation's Cities." *City Report.* City Observatory. http://cityobservatory.org/ynr/.

——. 2015a. "Surging City Center Job Growth." *City Report.* City Observatory. http://cityobservatory.org/city-center-jobs/

——. 2015b. "Less in Common." *City Report.* City Observatory. http://cityobservatory.org/less-in-common/

Cracknell, John. 2003. "TransMilenio Busway-Based Mass Transit, Bogotá, Colombia." WorldBank.http://siteresources.worldbank.org/INTURBANTRANSPORT/Resources/Factsheet-TransMilenio.pdf.

Creech, Brett. 2016. "Bureau of Labor Statistics: Beyond the Numbers, Pricing and Spending." United States Department of Labor. http://www.bls.gov/opub/btn/volume-5/pdf/expenditures-on-celluar-phone-services-have-increased-significantly-since-2007.pdf.

CSRwire. 2016. "CSR Profile of B Lab." http://www.csrwire.com/members/9730-B-Lab.

Dahmann, Nicholas, Jennifer Wolch, Pascale Joassart-Marcelli, Kim Reynolds, and Michael Jerrett. 2010. "The Active City? Disparities in Provision of Urban Public Recreation Resources." *Health and Place* 16 (2): 431–445. doi:10.1016/j.healthplace.2009.11.005.

Daley, Dorothy M., Elaine B. Sharp, and Jungah Bae. 2013. "Understanding City Engagement in Community-Focused Sustainability Initiatives." *Cityscape* 15 (1): 143–161.

Daniel, Robert, and Mark Render. 2003. "From Mule Tracks to Light Rail Transit Tracks." Presented at Transportation Research Circular E-C058: 9th National Light Rail Transit Conference. Presented in Portland, OR. November 16–18. http://onlinepubs.trb.org/onlinepubs/circulars/ec058/15_04_Daniel.pdf.

Davenport, Coral. 2016. "As Wind Power Lifts Wyoming's Fortunes, Coal Miners Are Left in the Dust." *New York Times,* June 19. http://www.nytimes.com/2016/06/20/us/as-wind-power-lifts-wyomings-fortunes-coal-miners-are-left-in-the-dust.html.

Davenport, Coral, and Jack Ewing. 2015. "VW Is Said to Cheat on Diesel Emissions; U.S. to Order Big Recall." *New York Times,* September 18. http://www.nytimes.com/2015/09/19/business/volkswagen-is-ordered-to-recall-nearly-500000-vehicles-over-emissions-software.html?_r=0.

Davis, Benjamin, and Phineas Baxandall. 2013. *Transportation in Transition: A Look at Changing Travel Patterns in America's Biggest Cities.* Washington, DC: U.S. PIRG Education Fund.

Department of Sanitation, City of New York (DSNY). 2006. "Final Comprehensive Solid Waste Management Plan."https://www1.nyc.gov/assets/dsny/docs/about_swmp_exec_summary_0815.pdf.

——. 2016. "Food Scraps + Yard Waste for Businesses." http://www1.nyc.gov/assets/dsny/zerowaste/businesses/food-scraps-and-yard-waste.shtml.

De Sousa, Christopher A. 2003. "Turning Brownfields into Green Space in the City of Toronto." *Landscape and Urban Planning* 62 (4): 181–198.

Deutsche Asset Management Center for Sustainable Finance. 2016. "Sustainable Finance Report." Deutsche Bank AG. https://institutional.deutscheam.com/content/_media /K15086_Sustainable_Finance_Report_RZ_160219.pdf.

De Vynck, Gerrit. 2016. "Vancouver Calls Itself Silicon Valley North—So Why No Uber?" *Financial Post,* July 8. http://business.financialpost.com/news/vancouver -calls-itself-silicon-valley-north-so-why-no-uber.

Dietzsch, Anna. 2015. "How Can Local Design Impact Large Infrastructure Plans and Projects?" *The Nature of Cities,* July 21. http://www.thenatureofcities.com/2015/06/21 /how-can-local-design-impact-large-infrastructure-plans-and-projects/.

Dilworth, Richardson, Robert Stokes, Rachel Weinberger, and Sabrina Spatari. 2011. "The Place of Planning in Sustainability Metrics for Public Works: Lessons from the Philadelphia Region." *Public Works Management & Policy* 16 (1): 20–39. doi:10.1177/1087724X10383891

Dimitri, Carolyn, Anne Effland, and Neilson Conklin. 2005. *The 20th Century Transformation of U.S. Agriculture and Farm Policy.* Washington, DC: US Department of Agriculture, Economic Research Service.

Dingman, Shane. 2015. "A Billionaire on Paper, Airbnb Co-founder Feels 'Great Responsibility' to Do Good." *Globe and Mail,* December 17. http://www.theglobeandmail .com/report-on-business/careers/careers-leadership/a-billionaire-on-paper -airbnb-co-founder-feels-great-responsibility-to-do-good/article27825035/.

Dittrich-Wesbuer, Andrea, Stefanie Föbker, and Frank Osterhage. 2008. "Demographic Change and Migration in City Regions: Results from Two German Case Studies." *ZAMM- Journal of Applied Mathematics and Mechanics* 33 (October): 315–350. doi:10.1007/s12523-009-0019-0.

Dunlap, Riley E. 2016. "Americans Believe 2015 Was Record-Warm, but Split on Why." Gallup, March 28. http://www.gallup.com/poll/190319/americans-believe-2015-record-warm -split-why.aspx?g_source=ENVIRONMENT&g_medium=topic&g_campaign =tiles.

DuPuis, Nicole, and Brooks Rainwater. 2015. "Shifting Perceptions of Collaborative Consumption." National League of Cities. https://seoulsolution.kr/sites/default/files /gettoknowus/Brief%20-%20Shifting%20Perceptions%20of%20Collaborative%20 Consumption2015.pdf.

Dwyer, Jim. 2016a. "De Blasio Adding Money for Water Tunnel in Brooklyn and Queens." *New York Times,* April 5. http://www.nytimes.com/2016/04/06/nyregion /de-blasio-postpones-work-on-crucial-water-tunnel.html.

——. 2016b. "De Blasio Postpones Work on Crucial Water Tunnel." *New York Times,* April 5. http://www.nytimes.com/2016/04/06/nyregion/de-blasio-postpones-work -on-crucial-water-tunnel.html.

Earth Institute. 2001. *Life After Fresh Kills: Moving Beyond New York City's Current Waste Management Plan.* New York: Columbia University's Earth Institute.

Economist. 2013. "The Rise of the Sharing Economy." *Economist,* March 9. http://www .economist.com/news/leaders/21573104-internet-everything-hire-rise-sharing -economy.

Environment Bureau. 2013. "Hong Kong: Blueprint for Sustainable Use of Resources (2013-2022)." http://www.enb.gov.hk/en/files/WastePlan-E.pdf.

EPD (Environmental Protection Department). 2006. "Waste Reduction Framework Plan: Executive Summary." http://www.epd.gov.hk/epd/english/environmentinhk /waste/prob_solutions/wrfp_doc.html.

——. 2013. "Monitoring of Solid Waste in Hong Kong: Waste Statistics for 2013." https:// www.wastereduction.gov.hk/sites/default/files/msw2013.pdf.

——. 2014. "Sustainable Use of Resources: Five-Pronged Approach." http://www.epd .gov.hk/epd/english/environmentinhk/waste/waste_maincontent.html.

——. 2016a. "Problems and Solutions: Implementation of Waste Disposal Plan." http:// www.epd.gov.hk/epd/english/environmentinhk/waste/prob_solutions/iwdp.html.

——. 2016b. "Recycling Fund." http://www.recyclingfund.hk/en/overview.php.

——. 2016c. "Source Separation of Domestic Waste: Achievements." https://www .wastereduction.gov.hk/en/household/source_achievements.htm.

Ernst, J. 2007. Factors Associated with K–12 Teachers' Use of Environment-Based Education. *Journal of Environmental Education* 38 (3): 15–32.

FAO (Food and Agriculture Organization). 2008. "An Introduction to the Basic Concepts of Food Security." http://www.fao.org/docrep/013/al936e/al936e00.pdf.

Federico, Carmela, and Jaimie Cloud. 2009. "Kindergarten Through Twelfth Grade Education: Fragmentary Progress in Equipping Students to Think and Act in a Challenging World, by Federico and Cloud." In *Agenda for a Sustainable America*, ed. John Dernbach, 109–128. Washington, DC: Environmental Law Institute.

Firestine, Theresa. 2016. "Bike-Share Stations in the United States." U.S. Department of Transportation, Bureau of Transportation Statistics. http://www.rita.dot.gov/bts /sites/rita.dot.gov.bts/files/publications/bts_technical_report/april_2016.

Fisher, Aaron, Sarah Zients, and Garrett Donnelly. 2015. "Checking on the Manhattanville Community Benefits Agreement, Six Years Later." *Columbia Spectator,* March 25. http://features.columbiaspectator.com/eye/2015/03/25/ties-that-bind/.

Flegenheimer, Matt. 2013. "After Decades, a Water Tunnel Can Now Serve All of Manhattan." *New York Times,* October 16. http://www.nytimes.com/2013/10/17/nyregion/new -water-tunnel-can-provide-water-for-all-of-manhattan.html?_r=1.

Frieling, Benjamin, and Nimrod Levy. n.d. "Yafo St. Traffic Flow & Air Pollution Levels." Jerusalem Municipality Environmental Department. http://www.polisnetwork.eu /uploads/Modules/PublicDocuments/air-pollution-study-on-jerusalem-light-rail- impact-(presentation).pdf.

Friends of the High Line. 2016. "About." http://www.thehighline.org/about.

Frizell, Sam. 2014. "The New American Dream Is Living in a City, Not Owning a House in the Suburbs." *Time,* April 25. http://time.com/72281/american-housing/.

FTA (Federal Transit Administration). 2016. "The National Transit Database." https:// www.transit.dot.gov/ntd.

Fuchs, Ester R. 2012. "Governing the Twenty-First-Century City." *Journal of International Affairs* 65 (2): 43–56. https://www.questia.com/read/1G1-287956972/governing -the-twenty-first-century-city.

Garcia, Miriam Pina. 2016. "Strong Measures to Combat Unlicensed Tourist Accommodation." *El Digital Barcelona,* June 28. http://eldigital.barcelona.cat/en/strong-measures-to-combat-unlicensed-tourist-accommodation_367135.html.

Garmendia, Maddi, Cecilia Ribalaygua, and José María Ureña. 2012. "High Speed Rail: Implication for Cities." *Cities* 29 (Suppl 2): S26–S31. http://www.sciencedirect.com/science/article/pii/S0264275112001011.

Garnett, Tara, Michael C. Appleby, Andrew Balmford, Ian J. Bateman, Tim G. Benton, Phil Bloomer, Barbara Burlingame, et al. 2013. "Sustainable Intensification in Agriculture: Premises and Policies." *Science* 341 (6141): 33–34. doi:10.1126/science.1234485.

Geron, Tomio. 2013. "Airbnb and the Unstoppable Rise of the Share Economy." *Forbes,* January 23. http://www.forbes.com/sites/tomiogeron/2013/01/23/airbnb-and-the-unstoppable-rise-of-the-share-economy/#78a089e16790.

Gillis, Justin. 2014. "Restored Forests Breathe Life Into Efforts Against Climate Change." *New York Times,* December 23. http://www.nytimes.com/2014/12/24/science/earth/restored-forests-are-making-inroads-against-climate-change-.html?_r=0.

Global Strategy Group. 2014. "Rockefeller Millennials Survey." http://t4america.org/wp-content/uploads/2014/04/RF-Millennials-Survey-Topline.pdf.

Global Times. 2014a. "Beijing Grapples with Overflowing Garbage Problems." *Global Times,* December 16. http://www.globaltimes.cn/content/897061.shtml.

——. 2014b. "Beijing Population Reaches 21 Million." *Global Times,* January 23. http://www.globaltimes.cn/content/839072.shtml.

Goodman, J. David. 2016. "5¢ Fee on Plastic Bags Is Approved by New York City Council." *New York Times,* May 5. http://www.nytimes.com/2016/05/06/nyregion/new-york-city-council-backs-5-cent-fee-on-plastic-bags.html?_r=0.

Goodyear, Sheena. 2015. "Uber vs. the World: How Cities Are Dealing with Ride-Hailing Technology." *CBC News,* October 2. http://www.cbc.ca/news/business/uber-versus-the-world-1.3252096.

Goulder, Lawrence H., and Robert N. Stavins. 2011. "Challenges from State-Federal Interactions in US Climate Change Policy." *American Economic Review* 101 (3): 253–257. doi:10.1257/aer.101.3.253.

Governing Magazine. 2015. "2015's Top Legislative Issues to Watch." *Governing Magazine,* January. http://www.governing.com/topics/politics/gov-issues-to-watch-2015.html.

——. 2016. "Public Transit Agency Ridership Statistics." Governing Data. http://www.governing.com/gov-data/transportation-infrastructure/public-transportation-agency-ridership-statistics-cities-metro-areas.html.

GovHK. 2016. "Waste Reduction & Separation." http://www.gov.hk/en/residents/environment/public/green/wastereduction.htm.

GreenBiz. 2016. "State of the Profession 2016." https://www.greenbiz.com/report/state-profession-2016.

GRI (Global Reporting Initiative). 2015. "G4 Sustainability Reporting Guidelines." https://www.globalreporting.org/resourcelibrary/GRIG4-Part1-Reporting-Principles-and-Standard-Disclosures.pdf.

Griffin, Jack, and Jim Merrihue. 2013. "CHP as a Reliable Energy Model—A Case Study from NYU." NYU and SourceOne, A Veolia Energy Company. http://www.districtenergy .org/assets/pdfs/2013AnnualConference/Monday/B3.2GRIFFINBRADLEYIDEA -Presentation-June-2013FINAL052413.pdf.

Griswold, Alison. 2015. "Uber's Siege on San Antonio." *Slate*, October 16. http://www .slate.com/articles/business/moneybox/2015/10/uber_returns_to_san_antonio_after _throwing_a_massive_tantrum_over_regulations.html.

Haanaes, Knut, Balu Balagopal, David Arthur, Ming Teck Kong, Ingrid Velken, Nina Kruschwitz, and Michael S. Hopkins. 2011. "First Look: The Second Annual Sustainability & Innovation Survey." *MIT Sloan Management Review* 52 (2): 77–83.

Hakim, Danny, Aaron Kessler, and Jack Ewing. 2016. "As Volkswagen Pushed to Be No. 1, Ambitions Fueled a Scandal." *New York Times*, September 26. http://www .nytimes.com/2015/09/27/business/as-vw-pushed-to-be-no-1-ambitions-fueled -a-scandal.html?hp&action=click&pgtype=Homepage&module=first-column -region&region=top-news&WT.nav=top-news.

Hamari, Juho, Mimmi Sjöklint, and Antti Ukkonen. 2016. "The Sharing Economy: Why People Participate in Collaborative Consumption." *Journal of the Association for Information Science and Technology* 67 (9): 2047–2059. doi:10.1002/asi.23552.

Hardcastle, Jessica Lyons. 2016. "Dow, 3M, Target Among Major Companies Advancing Circular Economy Principles." *Environmental Leader,* June 16. http://www .environmentalleader.com/2016/06/16/dow-3m-target-among-major-companies -advancing-circular-economy-principles/.

Harpaz, Joe. 2014 "Airbnb Disrupts Hotel Economy, Sends Regulators Scrambling." *Forbes*, May 7. http://www.forbes.com/sites/joeharpaz/2014/05/07/airbnb-disrupts -hotel-economy-sends-regulators-scrambling/#6f6b8a80726d.

——. 2016. "Airbnb to Big City Mayors: 'Take Our Tax Dollars, Please!'" *Forbes*, March 28. http://www.forbes.com/sites/joeharpaz/2016/03/28/airbnb-to-big-city-mayors-take -our-tax-dollars-please/#21d46614916d.

Harvey, Erika. 2015. "Turkeys on the High Line." Friends of the High Line, November 25. http://www.thehighline.org/blog/2015/11/25/turkeys-on-the-high-line.

Haugner, Mathias. 2015. "Following Fukushima Disaster, Japan Is Building Its First Microgrid Community." Team Gemini, July 21. http://teamgemini.us/following -fukushima-disaster-japan-is-building-its-first-microgrid-community/.

Herman, Pierre. 2014. "The AirBnB Phenomenon: What's the Impact on Cities?" Sustainable Cities Collective, July 24. http://www.sustainablecitiescollective.com/big-city /294081/airbnb-phenomenon-what-s-impact-cities.

Heyes, Anthony, Thomas P. Lyon, and Steve Martin. 2016. *Salience Games; Keeping Environmental Issues In (and Out) of the Public Eye*. Ross School of Business Working Paper, No. 1318. Ann Arbor: Ross School of Business, University of Michigan.

Hirshon, Lauren, Morgan Jones, Dana Levin, Kathryn McCarthy, Benjamin Morano, Sarah Simon, and Brooks Rainwater. 2015. "Cities, the Sharing Economy and What's Next." National League of Cities. http://www.nlc.org/Documents/Find%20City%20

Solutions/City-Solutions-and-Applied-Research/Report%20-%20%20Cities%20
the%20Sharing%20Economy%20and%20Whats%20Next%20final.pdf.

Hoogmartens, Rob., Maarten Dubois, and Steven Van Passel. 2016. "Identifying the Interaction Between Landfill Taxes and NIMBY." *Legal Aspects of Sustainable Development*,
497–509. Basel: Springer International Publishing. doi:10.1007/978-3-319-26021-1_24.

Huber, Joseph. 2000. "Towards Industrial Ecology: Sustainable Development as a Concept of Ecological Modernization." *Journal of Environmental Policy & Planning* 2:
269–285. doi:10.1080/714038561

Hutchinson, Alex. 2011. "TransMilenio: The Good, the Bus and the Ugly." *The City Fix*,
July 14. http://thecityfix.com/blog/transmilenio-the-good-the-bus-and-the-ugly/.

ICCT (International Council of Clean Transportation). 2015. "Supporting the Electric
Vehicle Market in U.S. Cities." http://www.theicct.org/sites/default/files/publications
/SupportEVsUScities_201510.pdf.

ICLEI Local Governments for Sustainability. 2016. "Sustainable City." http://www.iclei
.org/activities/agendas/sustainable-city.html.

IEA (International Energy Agency). 2016. "Global EV Outlook 2016." http://www.iea
.org/publications/freepublications/publication/Global_EV_Outlook_2016.pdf.

IRENA (International Renewable Energy Agency). 2015. "Battery Storage for Renewables:
Market Status and Technology Outlook." http://www.irena.org/documentdownloads
/publications/irena_battery_storage_report_2015.pdf.

Isaac, Mike. 2016. "How Uber Lost More than $1 Billion in the First Half of 2016." *New
York Times*, August 25. http://www.nytimes.com/2016/08/26/technology/how-uber
-lost-more-than-1-billion-in-the-first-half-of-2016.html.

ITDP (Institute for Transportation & Development Policy). 2007. "Bus Rapid Transit
Planning Guide." https://www.itdp.org/wp-content/uploads/2014/07/54.-Bus-Rapid
-Transit-Guide-PartIII-2007-09.pdf.

JICA. 2015. "Connect with the World and Revitalize Community" Higashimatsushima City in Cooperation with JICA. http://www.jica.go.jp/english/publications
/brochures/c8hovm0o0000k9ko-att/higashi-matsushima_en.pdf.

Jin, Qingxu., Carl Spevacek, Nasreddine El-Dehaibi, and Whitney Johnson. 2016. "Uber
and the Sharing Economy: Global Market Expansion and Reception." Ann Arbor:
William Davidson Institute Publishing, University of Michigan.

Jin, Tony. 2012. "China High Speed Train Development and Investment." *China Perspective*,
December 27. http://www.thechinaperspective.com/articles/chinahighspeedr-9905/.

Jinran, Zheng. 2015. "Three More Incinerators to Burn Mounting Waste." *China Daily*,
January 29. http://www.chinadaily.com.cn/china/2015-01/29/content_19435675.htm.

JLL Staff Reporter. 2016. "Moving at High Speed: Connecting China's Business Clusters." *JLL Real Views*, February 23. http://www.jllrealviews.com/economy/moving
-high-speed-connecting-chinas-business-clusters/.

Jo, Jin Ho, and Robert Laurence Martin. 2015. "Addressing Sustainability Issues in Complex Urban Systems Using a Sustainable Energy Plan." *International Journal of Sustainable Building Technology and Urban Development* 3 (4): 234–241. doi:10.1080/20
93761X.2012.745804.

Jolly, David. 2015. "Despite Push for Cleaner Cars, Sheer Numbers Could Work Against Climate Benefits." *New York Times,* December 7. http://www.nytimes.com/2015/12/08 /business/energy-environment/despite-push-for-cleaner-cars-sheer-numbers -could-work-against-climate-benefits.html?rref=collection%2Ftimestopic%2FElect ric%20and%20Hybrid%20Vehicles.

Jonas, Alexandra. 2016. "Share and Share Dislike: The Rise of Uber and Airbnb and How New York City Should Play Nice." *Journal of Law and Policy* 24 (1): 205.

Jones, Jeffrey. 2015. "In U.S., Concern About Environmental Threats Eases." Gallup, March 25. http://www.gallup.com/poll/182105/concern-environmental-threats-eases.aspx.

Kaneko, Kenji. 2015. "Sekisui House to Build Microgrid in Miyagi." *Japan Today,* March 26. http://jp.b2.mk/news/?newsid=gSW.

Karsten, Matthew. 2016. "What Are The Most Important Religious Sites in Jerusalem?" *Expert Vagabond,* May 26. http://expertvagabond.com/religious-sites-jerusalem/.

Katz, Bruce, and Jennifer Bradley. 2013. *Metropolitan Revolution: How Cities and Metros Are Fixing Our Broken Politics and Fragile Economy.* Washington, DC: Brookings Institution Press.

Katz, Lawrence F., and Alan B. Krueger. 2016. *The Rise and Nature of Alternative Work Arrangements in the United States, 1995-2015.* NBER Working Paper. http:// krueger.princeton.edu/sites/default/files/akrueger/files/katz_krueger_cws_-_ march_29_20165.pdf?version=meter+at+11&module=meter-Links&pgtype=article &contentId=&mediaId=&referrer=https%3A%2F%2Fwww.google.com%2F&priorit y=true&action=click&contentCollection=meter-links-click.

Kiernan, Paul. 2016. "Mining Dams Grow to Colossal Heights, and So Do the Risks." *Wall Street Journal,* April 5. http://www.wsj.com/articles/brazils-samarco-disaster -mining-dams-grow-to-colossal-heights-and-so-do-the-risks-1459782411.

Ki-moon, Ban. 2011. "A Vision Statement by Ban Ki-moon Secretary-General of the United Nations: Sustainable Energy for All." United Nations. http://www.se4all.org /sites/default/files/l/2013/09/SG_Sustainable_Energy_for_All_vision_final_clean.pdf.

Kirkland, Rik. 2014. "Interview with Brian Cheskey." McKinsey & Company. http:// www.mckinsey.com/industries/travel-transport-and-logistics/our-insights/the -future-of-airbnb-in-cities.

Korkki, Phyllis. 2014. "The Ripple Effects of Rising Student Debt." *New York Times,* May 24. http://www.nytimes.com/2014/05/25/business/the-ripple-effects-of-rising-student -debt.html.

Kumparak, Greg, Matt Burns, and Anna Escher. 2015. "A Brief History of Tesla." *Tech Crunch,* July 28. https://techcrunch.com/gallery/a-brief-history-of-tesla/slide/1/.

Kunkle, Fredrick. 2016. Uber Recognizes First Drivers Association in New York City. *Washington Post,* May 10. https://www.washingtonpost.com/news/tripping/wp/2016 /05/10/uber-recognizes-first-drivers-association-in-new-york-city/.

Lavrinc, Damon. 2013. "5 Cities Buy More Than Half of All Electric Vehicles in U.S." *Wired,* August 19. http://www.wired.com/2013/08/five-cities-ev/.

LeBar, Matthew. 2014. "MOOCs—Completion Is Not Important." *Forbes,* September 16. http:// www.forbes.com/sites/ccap/2014/09/16/moocs-finishing-is-not-the-important-part/.

Lee, Andrew C. K., and Ravi Maheswaran. 2010. "The Health Benefits of Urban Green Spaces: A Review of the Evidence." *Journal of Public Health* 33 (2): 212–222. doi:10.1093/pubmed/fdq068.

Lewis, Roger K. 2013. "Canal Park's Many Creative Ideas." *Washington Post*, March 15. https://www.washingtonpost.com/realestate/canal-parks-many-creative-ideas/2013/03/14/ea991520-7c8b-11e2-a044-676856536b40_story.html.

Loucks, Daniel P. 2008. "Water Resource Management Models." *Bridge* 38 (3). https://www.nae.edu/File.aspx?id=7423.

Loucks, Daniel P., and Eelco Van Beek. 2005. "Urban Water Systems." In *Water Resources Systems Planning and Management: An Introduction to Methods, Models and Applications*, 427–460. Paris: UNESCO.

Lowe, Ernest A., and Laurence K. Evans. 1995. "Industrial Ecology and Industrial Ecosystems." *Journal of Cleaner Production* 3 (1–2): 47–53. doi:10.1016/0959-6526(95)00045-G.

Lubber, Mindy. 2015. "Why Corporate America is Supporting EPA's Clean Power Plan." Ceres, August 3. https://www.ceres.org/press/blog-posts/why-corporate-america-is-supporting-epas-clean-power-plan.

Lustgarten, Abrahm. 2016. "Unplugging the Colorado River." *New York Times*, May 20. http://www.nytimes.com/2016/05/22/opinion/unplugging-the-colorado-river.html.

Lutz, Ashley. 2015. "An Emerging American Trend Is Terrifying News for Macy's, Kohl's, and JC Penney." *Business Insider,* May 14. http://www.businessinsider.com/consumers-spending-less-on-apparel-2015-5.

MacBride, Samantha. 2013. "San Francisco's Famous 80% Waste Diversion Rate: Anatomy of an Exemplar." *Discard Studies*, December 6. https://discardstudies.com/2013/12/06/san-franciscos-famous-80-waste-diversion-rate-anatomy-of-an-exemplar/.

MacDonald, Jennifer. 2016. "Electric Vehicles to Be 35% of Global New Car Sales by 2040." *Bloomberg New Energy Finance,* February 25. http://about.bnef.com/press-releases/electric-vehicles-to-be-35-of-global-new-car-sales-by-2040/.

Maclaren, Virginia W. 1996. "Urban Sustainability Reporting." *Journal of the American Planning Association* 62 (2): 184–202. doi:10.1080/01944369608975684.

Marnay, Chris, Hirohisa Aki, Keiichi Hirose, Alexis Kwasinski, Saori Ogura, and Takao Shinji. 2015. "Japan's Pivot to Resilience: How Two Microgrids Fared After the 2011 Earthquake." *IEEE Power and Energy Magazine* 13 (3): 44–57. doi:10.1109/MPE.2015.2397333.

Martignoni, Jimena. 2009. "Una plaza que descontamina." *La Nacion,* May 6. http://www.lanacion.com.ar/1124247-una-plaza-que-descontamina.

Matzler, Kurt, Viktoria Veider, and Wolfgang Kathan. 2014. "Adapting to the Sharing Economy." *MIT Sloan Management Review* 56 (2). http://sloanreview.mit.edu/article/adapting-to-the-sharing-economy/.

Mawad, Marie, Helene Fouquet, and Henry Goldman. 2016. "City Mayors Worldwide Forge Alliance in Response to Airbnb, Uber." *Bloomberg Technology,* June 20. http://www.bloomberg.com/news/articles/2016-06-20/city-mayors-worldwide-forge-alliance-in-response-to-airbnb-uber.

McKinsey & Company. 2010. "How Companies Manage Sustainability: McKinsey Global Survey Results." http://www.mckinsey.com/business-functions/sustainability-and-resource-productivity/our-insights/how-companies-manage-sustainability-mckinsey-global-survey-results.

McNichol, Andrew L. 2015. "Airbnb: A (Soon to Be) Victim of Its Own Success." *Legal Professionals* (Fall): 8–10. http://www.kutakrock.com/files/News/3170900e-afb9-4c33-9731-59e07489781b/Presentation/NewsAttachment/13994e1d-512e-457d-888f-83e6f5acod71/AirBNB_Fall15@Law.pdf.

Meyer, J. 2016. "Uber's New York Win." *Forbes*, January 19. http://www.forbes.com/sites/jaredmeyer/2016/01/19/uber-deblasio-new-york-study/#399feoab2b20.

Miller, Dale T., and Deborah A. Prentice. 2016. "Changing Norms to Change Behavior." *Annual Review of Psychology* 67: 339–361. doi:10.1146/annurev-psych-010814-015013.

Milman, Oliver, and Jessica Glenza. 2016. "At Least 33 US Cities Used Water Testing 'Cheats' over Lead Concerns." *Guardian*, June 2. https://www.theguardian.com/environment/2016/jun/02/lead-water-testing-cheats-chicago-boston-philadelphia.

Minter, Adam. 2016. "Just Say No to High-Speed Rail." *Bloomberg View*, July 6. https://www.bloomberg.com/view/articles/2016-07-07/just-say-no-to-high-speed-rail.

Mitsui & Co. 2012. "Solar Power Project to Support Higashimatsushima City Reconstruction." Mitsui & Co, December 4. http://www.mitsui.com/jp/en/release/2012/1205111_6471.html.

Mixon, Bobbie. 2015. "Improving Electric Vehicle Sales May Require Solving Unique Chicken and Egg Problem." National Science Foundation, January 29. http://www.nsf.gov/discoveries/disc_summ.jsp?cntn_id=133947.

Montezuma, Ricardo. 2005. "The Transformation of Bogotá, Colombia, 1995-2000: Investing in Citizenship and Urban Mobility." *Global Urban Development Magazine* 1 (1): 1–10.

Morgan, Kevin. 2015. "Nourishing the City: The Rise of the Urban Food Question in the Global North." *Urban Studies* 52 (8): 1379–1394. doi:10.1177/0042098014534902.

Morgan Stanley. 2016. "New Bonds Make It Easy to Be Green." http://www.morganstanley.com/articles/new-green-bonds.

Morgan Stanley Institute for Sustainable Investing. 2015. "Sustainable Signals: The Individual Investor Perspective." https://www.morganstanley.com/sustainableinvesting/pdf/Sustainable_Signals.pdf.

Mota, Ricardo. 2015. "Can Jerusalem's Light Rail Pull the City's Past Into the Future?" *Planetizen*, December 5. http://www.planetizen.com/node/82530/can-jerusalems-light-rail-pull-city%E2%80%99s-past-future.

Movellan, Junko. 2015. "Born from Disaster, Japan Established First Microgrid Community." *Renewable Energy World*, May 18. http://www.renewableenergyworld.com/articles/2015/05/born-from-disaster-japans-first-microgrid-community-represents-future-of-energy.html.

National League of Cities. 2013. "Food Systems." http://www.sustainablecitiesinstitute.org/topics/food-systems.

National Science Foundation. 2009. "Building a Sustainable Energy Future." National Science Board. http://www.nsf.gov/nsb/publications/2009/comments_se_report.pdf.

National Wildlife Federation. 2016. "Stop Hard Rock Mines from Poisoning Our Water!" http://www.nwf.org/What-We-Do/Energy-and-Climate/Climate-and-Energy/Reduce-Fossil-Fuel-Reliance/Mining-Loopholes.aspx.

Neibauer, Michael. 2013. "Deal of the Year: The Impact of Canal Park and 200 Eye St. SE on the Capitol Riverfront." *Washington Business Journal,* April 26. http://www.bizjournals.com/washington/print-edition/2013/04/26/deal-of-the-year-the-impact-of-canal.html.

New Climate Economy. 2015. "Low-Carbon Cities Are a US$17 Trillion Opportunity Worldwide." Press release, November 8. http://newclimateeconomy.net/content/press-release-low-carbon-cities-are-us17-trillion-opportunity-worldwide.

Newport, Frank, and Andrew Dugan. 2015. "College-Educated Republicans Most Skeptical of Global Warming." Gallup, March 26. http://www.gallup.com/poll/182159/college-educated-republicans-skeptical-global-warming.aspx.

New York State. 2016. "Reforming the Energy Vision." https://www.ny.gov/programs/reforming-energy-vision-rev.

New York State Office of the Attorney General. 2014. "A.G. Schneiderman Releases Report Documenting Widespread Illegality Across Airbnb's NYC Listings; Site Dominated by Commercial Users." Press release, October 16. http://www.ag.ny.gov/press-release/ag-schneiderman-releases-report-documenting-widespread-illegality-across-airbnbs-nyc.

Nielsen Global Survey of Corporate Social Responsibility. 2014. "Doing Well by Doing Good: Increasingly, Consumers Care About Corporate Social Responsibility, but Does Concern Convert to Consumption?" Nielsen, June 17. http://www.nielsen.com/us/en/insights/reports/2014/doing-well-by-doing-good.html.

Nixon, Ron. 2015. "Obama's Power Africa Project Is Off to a Sputtering Start." *New York Times,* July 21. http://www.nytimes.com/2015/07/22/world/africa/obamas-power-africa-project-is-off-to-a-sputtering-start.html?_r=1.

NOAA (National Oceanic and Atmospheric Administration). 2016. "Billion-Dollar Weather and Climate Disasters: Table of Events." http://www.ncdc.noaa.gov/billions/events.

Nolte, Amina. 2016. "Political Infrastructure and the Politics of Infrastructure: The Jerusalem Light Rail." *City* 20 (3). doi:10.1080/13604813.2016.1169778.

Norris, Floyd. 2013. "Bloated Government? Federal Employment at 47-Year Low." *New York Times,* October 22. http://economix.blogs.nytimes.com/2013/10/22/bloated-government-federal-employment-at-47-year-low/?_r=1.

NYSERDA (New York State Energy Research and Development Authority). 2016a. "Clean Energy Fund." http://www.nyserda.ny.gov/About/Clean-Energy-Fund.

——. 2016b. "Microgrids 101." http://www.nyserda.ny.gov/All-Programs/Programs/NY-Prize/Microgrids-101.

NYU (New York University). 2010. "NYU Climate Action Plan." http://www.nyu.edu/nyu2031/nyuinnyc/sustainability/climate-commitments.php.

——. 2016a. "About NYU." http://www.nyu.edu/about.html.

——. 2016b. "Energy." http://www.nyu.edu/life/sustainability/areas-of-focus/energy .html.

Office of the New York City Comptroller. 2016. "New York City's Millennials in Recession and Recovery." Bureau of Budget. http://comptroller.nyc.gov/wp-content/uploads /documents/NYC_Millennials_In_Recession_and_Recovery.PDF.

O'Sullivan, Feargus. 2016. "Berlin Is Banning Most Vacation Apartment Rentals." *CityLab,* April 28. http://www.citylab.com/housing/2016/04/airbnb-rentals-berlin-vacation -apartment-law/480381/.

Overton, Thomas W. 2014. "New York City Cogeneration Plant, New York City." *Power Magazine,* September 1. http://www.powermag.com/new-york-university-cogeneration -plant-new-york-city/.

Pappano, Laura. 2012. "The Year of the MOOC." *New York Times,* November 2. http:// www.nytimes.com/2012/11/04/education/edlife/massive-open-online-courses-are -multiplying-at-a-rapid-pace.html.

Patterson, Brittany. 2015. "Electric Car Owners Struggle to Find a Charge." *Scientific American,* July 16. http://www.scientificamerican.com/article/electric-car-owners -struggle-to-find-a-charge/.

Pearsall, Hamil. 2010. "From Brown to Green? Assessing Social Vulnerability to Environmental Gentrification in New York City." *Environment and Planning C: Government and Policy* 28 (5): 872–886. doi:10.1068/c08126.

Pedersen, Martin C. 2009. "Float On." *Metropolis Magazine,* July. http://www.metropolismag .com/July-2009/Float-On/.

Perella, Maxine. 2016. "Can You Put a Price Tag on Purpose? How Patagonia Highlights Value Over Cost." *Sustainable Brands,* June 21. http://www.sustainablebrands .com/news_and_views/business_models/maxine_perella/can_you_put_price_tag _purpose_patagonia_highlights_val.

Pew Research Center. 2015. "Democrats Hold Advantages on Environment, Abortion, Education; Neither Party Has Edge on the Economy." July 23. http://www.people-press .org/2015/07/23/gops-favorability-rating-takes-a-negative-turn/7-23-2015-1-55-46-pm/.

——. 2016. "Millennials." http://www.pewresearch.org/topics/millennials/.

Philadelphia Water. 2016. "Green City, Clean Waters." http://www.phillywatersheds.org /what_were_doing/documents_and_data/cso_long_term_control_plan.

Planning Institute of Australia. 2006. "Sustainable Water Use Policy Position Statement." https://www.planning.org.au/documents/item/225.

Pompeu de Toledo, Roberto. 2016. "Who We Are." Grupo Abril. http://grupoabril.com .br/en/quem-somos/victor-civita/biografia.

Ponisio, Lauren C., and Claire Kremen. 2016. "System-Level Approach Needed to Evaluate the Transition to More Sustainable Agriculture." *Proceedings of the Royal Society of London. Series B, Biological Sciences* 283 (1824). doi:10.1098/rspb.2015.2913.

Porter, Leslie, and Adrian J. Parker. 1993. "Total Quality Management—The Critical Success Factors." *Total Quality Management* 4 (1): 13-22. doi:10.1080/09544129300000003.

Portney, Kent E. 2013. *Taking Sustainable Cities Seriously*. Cambridge, MA: MIT Press.

Posen, Hannah A. 2016. "Ridesharing in the Sharing Economy." *Iowa Law Review* 101 (1): 405–434.

Potts, Jason, Matthew Lynch, Ann Wilkings, Gabriel Huppé, Maxine Cunningham, and Vivek Voora. 2014. "The State of Sustainability Initiatives Review 2014: Standards and the Green Economy." International Institute for Sustainable Development (IISD) and the International Institute for Environment and Development (IIED). https://www .iisd.org/pdf/2014/ssi_2014.pdf.

PRNewswire. 2015. "New Book Uses Global News Tweets to Question if MOOCs Growing at 201% Are Dead." *Business Journals,* January 21. http://www.bizjournals.com /prnewswire/press_releases/2015/01/21/MN10411.

Prothero, Andrea, Susan Dobscha, Jim Freund, William E. Kilbourne, Michael Luchs, Lucie K. Ozanne, and John Thoegersen. 2011. "Sustainable Consumption: Opportunities for Consumer Research and Public Policy." *Journal of Public Policy & Marketing* 30 (1): 31–38. doi:10.1509/jppm.30.1.31.

PwC (PricewaterhouseCoopers). 2015. "The Sharing Economy." PwC Consumer Intelligence Series. https://www.pwc.com/us/en/technology/publications/assets/pwc -consumer-intelligence-series-the-sharing-economy.pdf.

Pyper, Julia. 2013. "Microgrids Become a Reality as Superstorm Sandy's Anniversary Nears." *E&E News Business,* September 11. http://www.eenews.net/stories/1059987045.

Quattrone, Giovanni, Davide Proserpio, Daniele Quercia, Licia Capra, and Mirco Musolesi. 2016. "Who Benefits from the "Sharing" Economy of Airbnb?" In *Proceedings of the 25th International ACM Conference on World Wide Web (WWW)*. Session 32: 1385–1394. April 11–15. Montreal, Quebec, Canada. Republic and Canton of Geneva, Switzerland: International World Wide Web Conferences Steering Committee.

Rabinovitch, Simon. 2013. "China's High-Speed Rail Gets Back on Track." *Washington Post*, January 16. https://www.washingtonpost.com/world/asia_pacific/chinas-high -speed-rail-gets-back-on-track/2013/01/16/db508eea-5ffd-11e2-a389-ee565c81c565 _story.html.

Railway Gazette. 2013. "Funding Approved for Jerusalem Light Rail Extension." *Railway Gazette*, June 28. http://www.railwaygazette.com/news/news/middle-east/single-view /view/funding-approved-for-jerusalem-light-rail-extension.html.

Ramsey, Mike, and Cassandra Sweet. 2016. "Tesla and SolarCity Agree to $2.6 Billion Deal." *Wall Street Journal,* August 1. http://www.wsj.com/articles/tesla-and-solarcity -agree-to-2-6-billion-merger-deal-1470050724.

Rauch, Daniel E., and David Schleicher. 2015. *Like Uber, but for Local Government Policy: The Future of Local Regulation of the "Shared Economy."* NYU Marron Institute of Urban Management, Working Paper No. 21. http://marroninstitute.nyu.edu /uploads/content/The_Future_of_Local_Regulation_of_the_Shared_Economy.pdf.

Rayle, Lisa, Susan Shaheen, Nelson Chan, Danielle Dai, and Robert Cervero. 2014. "App-Based, On-Demand Ride Services: Comparing Taxi and Ridesourcing Trips and

User Characteristics in San Francisco." University of California Transportation Center (UCTC) Working Paper. http://tsrc.berkeley.edu/sites/default/files/Ridesourcing WhitePaper_Nov2014Update.pdf.

Reegle. 2016. "Definition: Cogeneration." http://www.reegle.info/glossary/776.

Regenerative Leadership Institute. 2007. "What Is Sustainable Living?" RLI Blog, February 17. https://www.regenerative.com/sustainable-living.

Rein, Lisa. 2014. "Millennials Exit the Federal Workforce as Government Jobs Lose Their Allure." *Washington Post*, December 15. https://www.washingtonpost.com /politics/millennials-exit-the-federal-workforce-as-government-jobs-lose-their -allure/2014/12/15/ea3d4418-7fd4-11e4-9f38-95a187e4c1f7_story.html.

Remnick, Noah, and Rick Rojas. 2016. "Toxic Passaic River to Get $1.38 Billion Cleanup over 10 Years." *New York Times*, March 4. http://www.nytimes.com/2016/03/05 /nyregion/toxic-passaic-river-to-get-1-38-billion-cleanup-over-10-years.html? _r=0.

Richtel, Matt. 2016. "San Francisco, 'the Silicon Valley of Recycling.'" *New York Times*, March 25. http://www.nytimes.com/2016/03/29/science/san-francisco-the-silicon -valley-of-recycling.html?_r=0.

RobecoSAM. 2015. "Measuring Intangibles." http://www.robecosam.com/images /Measuring_Intangibles_CSA_methodology_e.pdf.

Ross, Julia. 2008. "A China Environmental Health Project Research Brief: Management of Municipal Solid Waste in Hong Kong and Taipei." Wilson Center, March 13. https://www.wilsoncenter.org/sites/default/files/taiwan_hk_trash_mar08.pdf.

Rucker, Angela. 2014. "Prizes Powering Off-Grid Solutions in Power Africa Countries." USAID, September/October. https://www.usaid.gov/news-information/frontlines /powertrade-africa/prizes-powering-grid-solutions-power-africa-countries.

Rudoren, Jodi. 2014. "In Divided Jerusalem, Rail Line for Arabs and Jews Is Among the Fractures." *New York Times*, July 13. http://www.nytimes.com/2014/07/14/world/middleeast /in-divided-jerusalem-rail-line-for-arabs-and-jews-is-among-the-fractures.html?_r=1.

Sacks, Danielle. 2011. "The Sharing Economy." *Fast Company* (155): 88.

Said, Carolyn. 2016. "The Airbnb Effect." *San Francisco Chronicle*, July 12. http://www .sfchronicle.com/airbnb-impact-san-francisco-2015/#1.

SASB (Sustainability Accounting Standards Board). 2013. "Conceptual Framework of the Sustainability Accounting Standards Board." http://www.sasb.org/wp-content /uploads/2013/10/SASB-Conceptual-Framework-Final-Formatted-10-22-13.pdf.

Satterthwaite, David. 1997. "Sustainable Cities or Cities That Contribute to Sustainable Development?" *Urban Studies* 34 (10): 1667–1691. doi:10.1080/0042098975394.

Schawbel, Dan. 2015. "Elite Daily and Millennial Branding Release Landmark Study on the Millennial Consumer." Millennial Branding, January 20. http://millennialbranding .com/2015/millennial-consumer-study/.

Schechner, Sam. 2015. "Paris Confronts Airbnb's Rapid Growth." *Wall Street Journal*, June 25. http://www.wsj.com/articles/SB12147335600370333763904581058032330315292.

Schoolman, Ethan D., Mike Shriberg, Sarah Schwimmer, and Marie Tysman. 2014. "Green Cities and Ivory Towers: How Do Higher Education Sustainability Initiatives Shape Millennials' Consumption Practices?" *Journal of Environmental Studies and Sciences* (November): 1–13. doi:10.1007/s13412-014-0190-z.

Schor, Juliet. 2014. "Debating the Sharing Economy." Great Transition Initiative. http://www.tellus.org/pub/Schor_Debating_the_Sharing_Economy.pdf.

Siegel, Karen R., Kai McKeever Bullard, Giuseppina Imperatore, Henry S. Kahn, Aryeh D. Stein, Mohammed K. Ali, and K. M. Narayan. 2016. "Association of Higher Consumption of Foods Derived from Subsidized Commodities with Adverse Cardiometabolic Risk Among US Adults." The JAMA Network, August. doi:10.1001/jamainternmed.2016.2410.

Singer, Thomas. 2015. "Driving Revenue Growth Through Sustainable Products and Services." Conference Board. https://www.conference-board.org/publications/publicationdetail.cfm?publicationid=2975.

Sivak, Michael, and Brandon Schoettle. 2012. "Update: Percentage of Young Persons with a Driver's License Continues to Drop." *Traffic Injury Prevention* 13 (4): 341. doi:10.1080/15389588.2012.696755.

Slee, Tom. 2016. "Airbnb Is Facing an Existential Expansion Problem." *Harvard Business Review,* July 11. https://hbr.org/2016/07/airbnb-is-facing-an-existential-expansion-problem.

Somaney, Jay. 2016. "Uber Lost More than $1.2 Billion in the First Half of 2016 Alone." *Forbes,* August 26. http://www.forbes.com/sites/jaysomaney/2016/08/26/uber-lost-more-than-1-2-billion-in-the-first-half-of-2016-alone/#7cf4df2c1b77.

Spegele, Brian. 2016. "China Includes Green Cap in Economic Blueprint." *Wall Street Journal,* March 5. http://www.wsj.com/articles/china-includes-green-cap-in-economic-blueprint-1457164553.

State of Washington. 2014. "What's Happening at Gasworks Park?" Department of Ecology press release, August. https://fortress.wa.gov/ecy/gsp/DocViewer.ashx?did=29840.

Statista. 2016. "Statistics and Facts About Electric Mobility." http://www.statista.com/topics/1010/electric-mobility/.

Steg, Linda, Siegwart Lindenberg, and Kees Keizer. 2015. "Intrinsic Motivation, Norms, and Environmental Behavior: The Dynamics of Overarching Goals." *International Review of Environmental and Resource Economics* 9: 179–207. doi:10.1561/101.00000077.

Stengel, Richard. 2009. "For American Consumers, a Responsibility Revolution." *Time,* September 10. http://content.time.com/time/magazine/article/0,9171,1921618,00.html.

Stevenson, Verity. 2015. "Uber Reduced Drinking and Driving Deaths in California, Study Suggests." *Globe and Mail,* August 3. http://www.theglobeandmail.com/news/british-columbia/uber-reduced-drinking-and-driving-deaths-in-california-study-suggests/article25820101/.

Stewart, James B. 2016. "Everyone Despises SolarCity Deal, Except Tesla Shareholders." *New York Times*, August 4. http://www.nytimes.com/2016/08/05/business/everyone -despises-solarcity-deal-except-tesla-shareholders.html?_r=0.

Stokols, Andrew. 2015. "How China Can Leverage High-Speed Rail for Compact Urban Development." CityFix, World Resources Institute, April 19. http://thecityfix.com /blog/how-china-can-leverage-high-speed-rail-for-compact-urban-development -andrew-stokols/.

Storper, Michael, and Michael Manville. 2006. "Behaviour, Preferences and Cities: Urban Theory and Urban Resurgence." *Urban Studies* 43: 1247–1274. doi:10.1080 /00420980600775642.

Story, Louise. 2007. "Lead Paint Prompts Mattel to Recall 967,000 Toys." *New York Times*, August 2. http://www.nytimes.com/2007/08/02/business/02toy.html.

Stuart, Tessa. 2016. "It's Not Just Flint: America Has a Major Lead-in-Water Problem." *Rolling Stone*, January 28. http://www.rollingstone.com/politics/news/its-not-just-flint -america-has-a-major-lead-in-water-problem-20160128.

Student Loan Hero. 2016. "A Look at the Shocking Student Loan Debt Statistics for 2016." https://studentloanhero.com/student-loan-debt-statistics-2016/.

Sustainable SITES Initiative. 2016. "Washington Canal Park." http://www.sustainablesites .org/washington-canal-park.

Sutter, John D. 2012. "Tech Companies Make Progress on 'Blood Phones' and 'Conflict Minerals.'" *CNN*, August 16. http://www.cnn.com/2012/08/16/tech/gaming-gadgets /congo-blood-phones-report/.

Tajima, Kayo. 2003. "New Estimates of the Demand for Urban Green Space: Implications for Valuing the Environmental Benefits of Boston's Big Dig Project." *Journal of Urban Affairs* 25 (5): 641–655. doi:10.1111/j.1467-9906.2003.00006.x.

Tarr, Joel A. 1984. "Water and Wastes: A Retrospective Assessment of Wastewater Technology in the United States, 1800-1932." *Technology and Culture* 25 (2): 226–263 . doi:10.2307/3104713.

The Cultural Landscape Foundation. 2016. "Gas Works Park." http://tclf.org/landscapes /gas-works-park.

The White House. 2013. "Power Africa Fact Sheet." Office of the Press Secretary, June 30. https://www.whitehouse.gov/the-press-office/2013/06/30/fact-sheet-power-africa.

The White House. 2016. "FACT SHEET: Obama Administration Announces Federal and Private Sector Actions to Accelerate Electric Vehicle Adoption in the United States." Office of the Press Secretary, July 21. https://www.whitehouse.gov/the-press -office/2016/07/21/fact-sheet-obama-administration-announces-federal-and -private-sector.

Times of Israel Staff. 2016. "Jerusalem Approves Second Light Rail Route." *Times of Israel*, January 27. http://www.timesofisrael.com/jerusalem-approves-new-light-rail -route/.

TransMilenio. 2014. "History." http://www.transmilenio.gov.co/en/articles/history-0.

Turner, Micheal, Chuck Kooshian, and Steve Winkelman. 2012. "Case Study: Colombia's Bus Rapid Transit (BRT) Development and Expansion." Center for Clean Air Policy. http://ccap.org/resource/case-study-colombias-bus-rapid-transit-brt-development-and-expansion/.

Uber Newsroom. 2015. "Uber Unable to Operate in San Antonio After April 1." March 31. https://newsroom.uber.com/us-texas/uber-to-shutter-operations-in-san-antonio-on-april-1/.

UITP Observatory of Automated Metros. 2015a. "Mobility in Cities Database." http://www.uitp.org/sites/default/files/cck-focus-papers-files/MCD_2015_synthesis_web_0.pdf.

——. 2015b. "World Metro Figures, Statistics Brief." http://www.uitp.org/sites/default/files/cck-focus-papers-files/UITP-Statistic%20Brief-Metro-A4-WEB_0.pdf.

Ulrich, Roger S. 1981. "Natural versus Urban Sciences: Some Psycho-physiological Effects." *Environment and Behavior* 13 (September): 523–556. doi:10.1177/0013916581135001.

UNEP (United Nations Environment Programme). 2000. "Agenda 21," chapter 28. http://www.unep.org/Documents/Default.asp?DocumentID=52.

——. 2011. "Visions for Change: Recommendations for Effective Policies on Sustainable Lifestyles." http://www.unep.fr/shared/publications/pdf/DTIx1321xPA-VisionsForChange%20report.pdf.

——. 2012. "Cities and Buildings." Sustainable Consumption and Production Branch. http://www.unep.org/SBCI/pdfs/Cities_and_Buildings-UNEP_DTIE_Initiatives_and_projects_hd.pdf.

Ungar, Laura. 2016. "Lead Taints Drinking Water in Hundreds Of Schools, Day Cares Across USA." *USA Today,* March 17. http://www.usatoday.com/story/news/nation/2016/03/17/drinking-water-lead-schools-day-cares/81220916/.

UN Global Compact. 2016. "Victor Civita Square." United Nations. https://www.unglobalcompact.org/system/attachments/603/original/CoB19586-92D4-47B8-802F-6B2BF1365827.doc?1262614054.

UN-Habitat (United Nations Human Settlements Programme). 2002. *Sustainable Urbanisation: Achieving Agenda 21.* Nairobi: UN-Habitat; London: Department for International Development.

United Nations. 2013. "Sustainable Development Challenges: Chapter III Towards Sustainable Cities." In *World Economic and Social Survey 2013,* 53–84. Department of Economic and Social Affairs, United Nations.

——. 2014. "World Urbanization Prospects 2014 Revision." Department of Economic and Social Affairs. http://esa.un.org/unpd/wup/highlights/wup2014-highlights.pdf.

United Nations Data. 2016. "City Population by Sex, City, and City Type." http://data.un.org/Data.aspx?d=POP&f=tableCode%3A240.

USADF (United States African Development Foundation). 2016. "Power Africa: The USADF Off Grid Energy Challenge." USDAF website. http://www.usadf.gov/power-africa2/.

USAID (United States Agency for International Development). 2014. "GE, USADF and USAID Announce 22 African Innovators as Winners of Power Africa Off-Grid Energy Challenge." Press release, September 25. https://www.usaid.gov/ethiopia /press-releases/ge-usadf-and-usaid-announce-22-african-innovators-winners-power.

——. 2016. "The Roadmap." https://www.usaid.gov/sites/default/files/documents/1860 /USAID_PA_Roadmap_April_2016_TAG_5080pt.pdf.

USDA (United States Department of Agriculture). 2015. *Trends in U.S. Local and Regional Food Systems*. Washington, DC: Economic Research Service, USDA. [Presented to Congress in January.]

U.S. DOE (United States Department of Energy). 2014. "2014 Renewable Energy Data Book." National Renewable Energy Laboratory. http://www.nrel.gov/docs/fy16osti /64720.pdf.

——. 2016. "Co-Optimization of Fuels & Engines for Tomorrow's Energy-efficient Vehicles." http://www.nrel.gov/docs/fy16osti/66146.pdf.

U.S. DOT (United States Department of Transportation). 2016. "Smart City Challenge." https://www.transportation.gov/smartcity.

U.S. EIA (United States Energy Information Administration). 2016. "Frequently Asked Questions." https://www.eia.gov/tools/faqs/faq.cfm?id=105&t=3.

U.S. EPA (United States Environmental Protection Agency). 2004. "Primer for Municipal Wastewater Treatment Systems." https://www.epa.gov/sites/production/files /2015-09/documents/primer.pdf.

——. 2014. "Enhancing Sustainable Communities with Green Infrastructure." https:// www.epa.gov/sites/production/files/2014-10/documents/green-infrastructure.pdf.

——. 2015. "Advancing Sustainable Materials Management: Facts and Figures". Office of Resource Conservation and Recovery. https://www.epa.gov/sites/production/files /2015-09/documents/2013_advncng_smm_rpt.pdf.

——. 2016. "Brownfields." https://www.epa.gov/brownfields.

USGBC (United States Green Building Council). 2009. "Washington Canal Park." http:// www.usgbc.org/projects/washington-canal-park-0?view=scorecard.

US PIRG Education Fund. 2014. "Millennials in Motion: Changing Travel Habits of Young Americans and the Implications for Public Policy." Frontier Group. http:// www.uspirg.org/sites/pirg/files/reports/Millennials%20in%20Motion%20USPIRG .pdf.

Van Daalen, Robin. 2014. "Airbnb to Collect Tourist Taxes in Amsterdam." *Wall Street Journal*, December 18. http://blogs.wsj.com/digits/2014/12/18/airbnb-to-collect -tourist-taxes-in-amsterdam/.

Vault. 2016. "2016 Vault Consulting 50." http://www.vault.com/company-rankings/consulting /vault-consulting-50/.

Walmart. 2015. "Walmart Marks Fulfillment of Key Global Responsibility Commitments." http://news.walmart.com/news-archive/2015/11/17/walmart-marks-fulfillment -of-key-global-responsibility-commitments.

Walsh, Bryan. 2013. "The Triple Whopper Environmental Impact of Global Meat Production." *Time,* December 16. http://science.time.com/2013/12/16/the-triple-whopper-environmental-impact-of-global-meat-production/.

Wang, Hao, and Chunmei Wang. 2013. "Municipal Solid Waste Management in Beijing: Characteristics and Challenges." *Waste Management & Research* 31 (1): 67–72. doi:10.1177/0734242X12468199.

Wang, XiaoHu, Christopher Hawkins, and Evan Berman. 2014. "Financing Sustainability and Stakeholder Engagement: Evidence from U.S. Cities." *Urban Affairs Review* 50 (6): 806–834. doi:10.1177/1078087414522388.

Wang, XiaoHu, Christopher V. Hawkins, Nick Lebredo, and Evan M. Berman. 2012. "Capacity to Sustain Sustainability: A Study of US Cities." *Public Administration Review* 72 (6): 841–853. doi:10.1111/j.1540-6210.2012.02566.x.

Wasserstein, Bernard, Stewart Henry Perowne, and Joshua Prawer. 2016. "Jerusalem." *Encyclopedia Britannica.* https://www.britannica.com/place/Jerusalem/Economy.

Wattles, Jackie. 2015. "Bill Gates Launches Multi Billion Dollar Clean Energy Fund." *CNN,* November 30. http://money.cnn.com/2015/11/29/news/economy/bill-gates-breakthrough-energy-coalition/.

Watts, Jonathan. 2012. "Beijing Introduces Recycling Banks That Pay Subway Credits for Bottles." *Guardian,* July 4. http://www.theguardian.com/environment/2012/jul/04/beijing-recycling-banks-subway-bottles.

Way, Thaïsa. 2013. "Landscapes of Industrial Excess: A Thick Sections Approach to Gas Works Park." *Journal of Landscape Architecture* 8 (1): 28–39. doi:10.1080/18626033.2013.798920.

Weisbrod, Glen, Derek Cutler, and Chandler Duncan. 2014. *Economic Impact of Public Transportation Investment.* Washington, DC: American Public transportation Association (APTA). http://www.apta.com/resources/reportsandpublications/Documents/Economic-Impact-Public-Transportation-Investment-APTA.pdf.

Wheeler, Candace. 2012. "Opening Weekend for Canal Park." *Washington Post,* November 16. https://www.washingtonpost.com/local/opening-weekend-for-canal-park/2012/11/16/fdf931a0-2f46-11e2-9f50-0308e1e75445_story.html#comments.

Wheeler, Keith A., and John M. Byrne. 2003. "K-12 Sustainability Education: Its Status and Where Higher Education Should Intervene." *Planning for Higher Education* 31 (3): 23–29.

Willmott, Don. 2015. "Devastated Town Rebuilds for Energy Self-Sufficiency." *The Huffington Post,* November 10. http://www.huffingtonpost.com/x-prize-foundation/devastated-town-rebuilds_b_8526918.html.

Winter, Mick. 2007. *Sustainable Living: For Home, Neighborhood and Community.* Napa, CA: Westsong Publishing.

Wittenberg, Ariel. 2016. "Fast-Charge Plus Do Not Fit All Electric Cars." *Scientific American,* August 1. http://www.scientificamerican.com/article/fast-charge-plugs-do-not-fit-all-electric-cars/.

Woo, Christian P., and Richard A. Bales. Forthcoming. "The Uber Million Dollar Question: Are Uber Drivers Employees or Independent Contractors?" *Mercer Law Review.*

World Bank. 2013a. "Global Waste on Pace to Triple by 2100." http://www.worldbank .org/en/news/feature/2013/10/30/global-waste-on-pace-to-triple.

——. 2013b. "Sustainable Cities Initiative." http://www.worldbank.org/en/region/eca/brief /sustainable-cities-initiative.

——. 2014a. "High-Speed Railways in China: A Look at Construction Costs." World Bank Paper. http://www.worldbank.org/en/news/press-release/2014/07/10/cost-of-high -speed-rail-in-china-one-third-lower-than-in-other-countries.

——. 2014b. *International Bank for Reconstruction and Development Project Appraisal Document on a Proposed Grant from the Global Environment Facility Trust.* Report No. PAD388. http://documents.worldbank.org/curated/en/748681468029344265 /pdf/PAD3880PAD0P12010Box385354B00OUO090.pdf.

——. 2016. "Introduction to Wastewater Treatment Processes." http://water.worldbank.org /shw-resource-guide/infrastructure/menu-technical-options/wastewater-treatment.

WHO (World Health Organization). 2015. "Drinking-water." http://www.who.int/mediacentre /factsheets/fs391/en/.

——. 2016. "Air Pollution Levels Rising in Many of the World's Poorest Cities." News release, May 12. http://www.who.int/mediacentre/news/releases/2016/air-pollution -rising/en/.

Wright, Lloyd. 2014. "Module 3b: Bus Rapid Transit." In *Sustainable Transport: A Sourcebook for Policy-makers in Developing Cities.* Institute for Transportation and Development Policy. https://www.itdp.org/wp-content/uploads/2014/07/Module-3b -Bus-Rapid-Transit.pdf.

Yang, Yukun, Ning Su, and Yulin Fang. 2016. "Didi, Kuaidi, and Uber in China." *Harvard Business Review,* April 12. https://hbr.org/product/didi-kuaidi-and-uber-in-china /W16194-PDF-ENG.

Yau, Yung. 2010. "Domestic Waste Recycling, Collective Action and Economic Incentive: The Case in Hong Kong." *Waste Management* 30: 2440–2447. doi:10.1016/j.wasman .2010.06.009.

Yavar, Javiera. 2014. "Arquitectura y Paisaje: Plaza Victor Civita, un museo abierto de Sustentabilidad por Levisky Arquitetos + Anna Dietzsch." *ArchDaily Colombia,* July 13. http://www.archdaily.co/co/02-376340/arquitectura-y-paisaje-plaza-victor -civita-un-museo-abierto-de-sustentabilidad-por-levisky-arquitetos-anna-dietzsch.

Yin, Juelin, Lixian Qian, and Anusorn Singhapakdi. 2016. "Sharing Sustainability: How Values and Ethics Matter in Consumers' Adoption of Public Bicycle-Sharing Scheme." *Journal of Business Ethics* (January). doi:10.1007/s10551-016-3043-8.

Ying, Zhu. 2014. "High Speed Railway's Impact on China." *MIRR Initiative White Book.* http://mir-initiative.com/white-book/zhu-ying/high-speed-railway%E2%80%99s -impact-on-shina.html.

Zahran, Sammy, Samuel Brody, Arnold Vedlitz, Himanshu Grover, and Caitlyn Miller. 2008. "Vulnerability and Capacity: Explaining Local Commitment to Climate-Change Policy." *Environment and Planning C: Government and Policy* 26, 544–562. doi:10.1068/c2g.

Zervas, Georgios., Davide Proserpio, and John W. Byers. 2016. *The Rise of the Sharing Economy: Estimating the Impact of Airbnb on the Hotel Industry.* Boston University School of Management Research Paper No. 2013-16. doi:10.2139/ssrn.2366898.

Zhang, Dong Qing, Soon Keat Tan, and Richard M. Gersberg. 2010. "Municipal Solid Waste Management in China: Status, Problems and Challenges." *Journal of Environmental Management* 91: 1623–1633. doi:10.1007/s10163-016-0509-9.

Zheng, Siqi, and Matthew Kahn. 2013. "China's Bullet Trains Facilitate Market Integration and Mitigate the Cost of Megacity Growth." *Proceedings of the National Academy of Sciences of the United States of America* 110 (14). doi:10.1073/pnas.1209247110.

Zhen-shan, Li, Yang Lei, Qu Xiao-Yan, and Sui Yu-mei. 2009. "Municipal Solid Waste Management in Beijing City." *Waste Management* 29: 2596–2599. doi:10.1016/j.wasman.2009.03.018.

INDEX